Quantum Ontology

Quantum Ontology
A GUIDE TO THE METAPHYSICS
OF QUANTUM MECHANICS

PETER J. LEWIS

UNIVERSITY PRESS

Oxford University Press is a department of the University of Oxford. It furthers the University's objective of excellence in research, scholarship, and education by publishing worldwide. Oxford is a registered trade mark of Oxford University Press in the UK and in certain other countries.

Published in the United States of America by Oxford University Press
198 Madison Avenue, New York, NY 10016, United States of America.

© Oxford University Press 2016

All rights reserved. No part of this publication may be reproduced, stored in a retrieval system, or transmitted, in any form or by any means, without the prior permission in writing of Oxford University Press, or as expressly permitted by law, by licence or under terms agreed with the appropriate reproduction rights organization. Inquiries concerning reproduction outside the scope of the above should be sent to the Rights Department, Oxford University Press, at the address above

You must not circulate this work in any other form
and you must impose this same condition on any acquirer.

Library of Congress Cataloging-in-Publication Data
Names: Lewis, Peter J., author.
Title: Quantum ontology : a guide to the metaphysics of quantum mechanics / Peter J. Lewis.
Description: New York, NY : Oxford University Press, 2016. |
Includes bibliographical references.
Identifiers: LCCN 2015039510| ISBN 9780190469818 (pbk. : alk. paper) | ISBN 9780190469825 (hardcover : alk. paper)
Subjects: LCSH: Quantum theory—Philosophy. | Metaphysics. | Ontology.
Classification: LCC QC174.12 .L49 2016 | DDC 530.12—dc23
LC record available at http://lccn.loc.gov/2015039510

9 8 7 6 5 4 3 2 1
Printed by Sheridan Books, Inc., United States of America

To my parents, Pat and John Lewis

CONTENTS

Preface xi
Introduction xiii

1 **Phenomena and Theory** 1
1.1 Interference 2
1.2 Entanglement 7
1.3 Matrix Mechanics 10
1.4 Wave Mechanics 17
1.5 Interpretation 22

2 **Realism** 25
2.1 Quantum Mechanics as Incomplete 27
2.2 No-Go Theorems 34
2.3 What Do the Theorems Prove? 38
2.4 Rescuing Realism 40
2.5 Conclusion 42

3 **Underdetermination** 44
3.1 The Measurement Problem 45
3.2 Spontaneous Collapse Theories 50
3.3 Hidden Variable Theories 55
3.4 The Many-Worlds Theory 59
3.5 Reducing the Alternatives 64
3.6 Conclusion 70

4 Indeterminacy   72
4.1 Textbook Indeterminacy   75
4.2 Radical Indeterminacy   79
4.3 Moderate Indeterminacy   86
4.4 Indeterminacy and Branching   97
4.5 Avoiding Indeterminacy   103
4.6 Conclusion   105

5 Causation   107
5.1 Locality   108
5.2 Particle Trajectories   111
5.3 Wave Packets   119
5.4 Collapses as Causes   122
5.5 Conclusion   126

6 Determinism   128
6.1 Uncertainty   129
6.2 Probability   135
6.3 Immortality   142
6.4 Free Will   145
6.5 Conclusion   149

7 Dimensions   151
7.1 Configuration Space   152
7.2 Three-Dimensionality as an Illusion   154
7.3 Adding Ontology   157
7.4 Interpreting the Wave Function   160
7.5 Conclusion   163

8 Parts and Wholes   165
8.1 The Case for Holism   166
8.2 Holism Reconsidered   168
8.3 The Case Against Individuals   173

8.4 Individuals Reconsidered   176
8.5 Conclusion   178

9 Six Quantum Worlds   179

Notes   183
References   194
Index   203

PREFACE

This book is intended as a guide to quantum mechanics for the philosophical consumer. Quantum mechanics is a rich source of metaphysical insight—about the nature of individuals and of properties, about supervenience and space and causation, about determinacy and determinism. This is not to say that physics holds all the answers. Quantum mechanics is notoriously slippery territory: Multiple interpretations of its basic mathematical formulation are possible, with very different implications about the nature of the world. In general, any metaphysical claim of the form "Quantum mechanics entails X" is likely to be false. But despite the lack of definitive answers, quantum mechanics is very important to metaphysics, because of the way it broadens the range of metaphysical possibilities, because of the way it challenges our classically tutored intuitions, and because of the way it reshapes metaphysical debates in surprising, empirically informed ways.

There are many things that this book is not. It is not a user's guide to quantum mechanics: Physicists need no help from philosophers like me, and the business of using quantum mechanics goes on largely independently of foundational metaphysical concerns. It is not a physics textbook: While I explain the bare bones of the theory and a couple of applications, the exposition of the physics is limited to what I need to get the metaphysical points across. There are plenty of good physics textbooks out there, some of which I mention along the way. It is not a polemic, or an argument in favor of one interpretation of quantum mechanics over the others, or an attempt to construct a new interpretation: There are plenty of good examples of these genres out there, too, many of which I mention.

While I have opinions about the various interpretations on offer, I try to remain even-handed for present purposes, and just lay out the options.

So if you are a philosopher (professional, student, or amateur) who wants to find out how quantum mechanics might be relevant to your metaphysical views, this book is for you. It is intended to be self-contained: I explain the physics, as well as the philosophy, along the way. I don't shy away from equations when necessary: Quantum mechanics is written in the language of mathematics, and to understand the theory you have to see for yourself what it says. But you don't need much math to understand the basics of quantum mechanics, and all the equations are fully explained, often with diagrams. Similarly, you don't need prior acquaintance with any particular metaphysical tradition—not least because the quantum-inspired debates are often quite distinct from the traditional ones.

I would like to acknowledge the generous support of the University of Miami in helping me to free up time to write this book: a Provost's Research Award in the summer of 2012, a Humanities Center Fellowship for 2012–2013, and a research leave under the College of Arts and Sciences Associate Professor Pilot Program during spring 2014.

I would also like to thank the many people who have helped me with this project. The 2012–2013 fellows of the University of Miami Humanities Center provided me with invaluable feedback about my attempts to make quantum mechanics accessible to the nonspecialist. Jenann Ismael gave insightful counsel about how to organize the material, as well as much-needed encouragement. The students in my *Philosophy and Quantum Mechanics* seminar in spring 2015 (David DiDomenico, Lou Enos, Iago Bozza Francisco, Wei Huang, Nihel Jhou, and Rina Tzinman) picked lots of holes in a draft of the book, as well as finding an embarrassing number of typos. Craig Callender, Roman Frigg, Jonathan Schaffer, and Eric Schwitzgebel gave me useful feedback on drafts of various chapters. Finally, my special thanks to Jeff Barrett, for teaching me how to think about and write about quantum mechanics, and for insightful comments on the whole manuscript.

This book would not have been possible without the encouragement, support, and philosophical acumen of my wife, Amie Thomasson. It would definitely have been possible without our daughters, Natalie and May, but it wouldn't have been nearly so much fun.

INTRODUCTION

Why is this book about quantum mechanics? That is, why, of all the scientific theories you might study, should metaphysicians be particularly interested in quantum mechanics, rather than string theory or molecular biology? This might seem to go without saying, for at least two reasons. First, it is tempting to think that quantum mechanics is interesting to metaphysicians because it is fundamental: It describes the basic structures of the physical world. Second, it is tempting to think that quantum mechanics is interesting to metaphysicians because it is revisionary: The descriptions of the world it provides are at odds with our metaphysical intuitions. But the first of these tempting answers is hard to defend, and the second, while reasonable, is apt to be somewhat misleading unless stated more carefully. So before we get started, it is worth trying to get a little clearer on the relationship between physics and metaphysics.

Metaphysicians have been interested in physics for as long as the two disciplines have been distinct (and before that, trivially so). And this seems quite natural, at least at first glance: Metaphysics is the branch of philosophy that deals with the fundamental structures of reality, and physics is the branch of science that does the same. Indeed, a commitment to the right combination of reductionism and naturalism might lead you to the conclusion that metaphysics is just physics, perhaps expressed in more philosopher-friendly terms. On that view, since quantum mechanics is a very important part of physics, it is ipso facto a very important part of metaphysics.

Most metaphysicians, though, don't accept this combination of reductionism and naturalism—perhaps none do. So the first-glance explanation for the interest of metaphysicians in physics would at least need to be carefully hedged, and I strongly suspect it is untenable. But fortunately we can bypass these concerns, since it is relatively straightforward to establish that understanding quantum mechanics is important for metaphysicians of virtually any stripe. The argument is this: Whatever one's philosophical commitments (pretty much), metaphysics is constrained by consistency with experience. Quantum phenomena are part of experience: The experiences of the physicist in the lab are no less constraints on metaphysics than the experiences of the philosopher playing with a piece of wax. So quantum mechanics is a constraint on metaphysics.

Of course, that doesn't entail that it's an *important* constraint. Poached-egg phenomena are part of experience, too, but though these phenomena are relevant to metaphysics (Lewis, 1979, 530), nobody should write a book on the metaphysical implications of poached eggs. One might suspect quantum mechanics of being even less central to metaphysics than poached eggs: Unless you are a professional physicist, one might think, you just don't run across experiences of the quantum world. But that would be a mistake. There are everyday phenomena that can only be explained in quantum mechanical terms—the light from your laser pointer, for example. Furthermore, quantum mechanics is not limited to the behavior of the very small, or of matter under certain extreme circumstances; it holds, if it holds at all, of all material objects all the time. Most of your experience of material objects may be roughly explicable in classical terms, but if modern physics is right, a more general explanation is in terms of quantum mechanics, since classical mechanics has been superseded by quantum mechanics and subsumed as a limiting case.[1] So your experience of material objects is always experience of quantum phenomena. Quantum mechanics is important to metaphysics because quantum phenomena, unlike poached-egg phenomena, are ubiquitous in our experience of the physical world.

I stand by this above argument, but put in such stark terms it runs the risk of applying either too broadly or too narrowly. Certainly quantum mechanics is relevant to the behavior of every object that we experience, but so are any number of other theories in physics (thermodynamics, optics, etc.) and chemistry (bonding theory, reaction kinetics, etc.). Any sufficiently general theory applying to material objects constitutes a ubiquitous constraint on our experience, and hence is relevant to metaphysics according to this argument. Perhaps some or all of these theories are reducible to quantum mechanics, but I don't want to tie my argument to any particular account of intertheoretic reduction. So the argument applies too broadly: It doesn't single out quantum mechanics as especially relevant to metaphysics.

Conversely, one might argue that, strictly speaking, quantum mechanics (the standard nonrelativistic theory outlined in Chapter 1) doesn't apply to any material objects, because it too has been superseded by quantum field theory and subsumed as a limiting case. Indeed, we know that even quantum field theory cannot be the final word, as it cannot be extended to gravitational phenomena. Quantum mechanics is false: At best it is approximately true within a certain range of application (Monton, 2013, 154). But in that case, quantum mechanics *isn't* the underlying explanation for the behavior of objects. So the argument applies too narrowly: Only the ultimate physical "theory of everything" (if there is such a thing) is relevant to metaphysics.

The response to both of these objections is basically the same: Quantum mechanics provides a level of physical theory at which a number of interrelated conceptual difficulties appear, and it is the responses to these conceptual difficulties that make quantum mechanics so metaphysically revisionary. So while it is true that many theories can be regarded as ubiquitous constraints on our experience, most of them are not terribly interesting from a metaphysical point of view.

Classical mechanics provides the most obvious case in point. Classical mechanics, like quantum mechanics, can be regarded as approximately true within a certain range of application, and indeed most of the observed behavior of macroscopic objects can be given classical mechanical explanations. But classical mechanics is not metaphysically revisionary; indeed, you might regard the description of the world provided by classical mechanics as the received metaphysical view. If philosophers wish to gesture at the physical reality underlying everyday phenomena, they typically do so in terms of a classical particle ontology—in terms of little billiard balls flying around according to Newton's laws of motion. A representative example is Merricks's analysis of ordinary objects: A statue, says Merricks, is nothing but "atoms arranged statuewise," where the atoms are "the atoms of physics, not Democritus" (Merricks, 2001, 3). This is not to single out Merricks: Such examples are commonplace. Admittedly, they are often accompanied by a disclaimer to the effect that claims about particles or atoms are "really placeholders for claims about whatever microscopic entities are actually down there" (Merricks, 2001, 3). But this disclaimer is itself couched in classical terms: There is no guarantee that physics will produce a fundamental ontology that is anything like microscopic entities arranged in three-dimensional space, and indeed quantum mechanics has been taken to show otherwise. Hence, a metaphysical exposition of classical mechanics would be otiose, since it would be merely telling metaphysicians what they intuitively assume, but a metaphysical exposition of quantum mechanics has the potential to reveal where those intuitions fail.

To an extent, every scientific theory (that is roughly empirically adequate) is metaphysically relevant, since every such theory rests on empirical phenomena, and every empirical phenomenon is a constraint on metaphysics. But most of these constraints are unsurprising, since they are already built into our standard metaphysical intuitions. What makes quantum mechanics exciting is that the phenomena on which it is based seem to *undermine* those intuitions. Note that it is an advantage of this phenomena-based motivation for the study of quantum mechanics that it is independent of whether quantum mechanics is a *fundamental* theory. Quantum mechanics, at least in the standard nonrelativistic form to be covered in this book, is surely not a fundamental or ultimate theory in any sense, but the argument developed here for its metaphysical interest does not depend on any such claim. A case can be made that evolutionary biology is also metaphysically revisionary, in that the phenomena of speciation cannot be captured by our intuitive species ontology (Okasha, 2002). So there is no "physics first" or reductionist agenda here: The metaphysical interest of quantum mechanics does not rest on its being an important part of modern *physics* per se. Rather, the claim is just that one particular set of empirical phenomena—those associated with quantum mechanics—provides a fertile area for metaphysical investigation.

So it doesn't matter that quantum mechanics isn't the final word in physical theory, and it doesn't matter that quantum mechanics doesn't reveal the ultimate constituents of matter. Perhaps those ultimate entities are the strings described by string theory; perhaps they are something else entirely. We don't really know, in large part because there are as yet no empirical results that bear on the matter. But what we do know is that whatever the ultimate entities are, they must exhibit the characteristic phenomena of quantum mechanics at the relevant level of description. At the level of electrons, atoms, and the like, we know from the results of extensive experimentation that quantum mechanics makes just the right predictions. It is these predictions that generate the metaphysical difficulties to be explored in this book. At the end of the day, it is quantum *phenomena* that are metaphysically problematic; these phenomena are well explored experimentally, and theoretical developments are not going to make them go away.

So much for the claim that quantum mechanics is interesting because it is fundamental. What about the claim that it is interesting because it is revisionary? As should be clear from the earlier discussion, I endorse this claim. But note the stress on the quantum phenomena rather than the theory of quantum mechanics. This is not accidental. The theory of quantum mechanics is notoriously difficult to interpret, to such an extent that several prominent physicists and philosophers have denied that it provides us with any description of the physical world at all. And even if one does take it as descriptive, there is no consensus about the nature of the description it provides us with. So it is certainly not straightforwardly true that quantum mechanics is metaphysically revisionary in the sense that it provides us with a theoretical description of the world that undermines our classically tutored intuitions.

Nevertheless, we can say this much: The quantum phenomena (as we shall see) are such that they cannot be accommodated within a classical world. So either we have to give up on the project of describing the world behind the phenomena at all (which I don't recommend), or we have to embark on the project of modifying our classical worldview to accommodate them. It is this latter project that takes up the bulk of the book. As will become clear, we can say quite confidently that quantum mechanics is metaphysically revisionary even if it is not clear what form the revisions should take.

There are two competing dangers to taking physics as a guide to metaphysics. The first danger is that we will convince ourselves that only the final, fundamental theory has anything definitive to tell us about ontology. In that case, we end up endorsing a kind of metaphysical skepticism: We can't know anything about physical ontology until the end of physics. This seems unwarranted: We can know a lot of things about rocks and organisms and molecules without knowing anything about their ultimate constituents. So there is no reason in principle why we should not know about the entities and processes involved in quantum phenomena without waiting for the fundamental theory in which they are ultimately grounded.[2]

The second danger is that we will take some particular account of the description of the world that quantum mechanics presents us with as definitive, and hence become overly confident of the metaphysical consequences of that description. The second danger points in the opposite direction to the first: We end up saying too much rather than too little. And this is not just an idle worry, since many of the metaphysical conclusions that have been claimed on the basis of quantum mechanics (as we shall see) do so on the basis of some particular interpretation of the quantum world, when other interpretations would entail different conclusions. The goal of this book is to navigate between these competing dangers: to say what can be said about the ontological implications of quantum mechanics, without overstating the case.

The first three chapters set up the general framework for drawing metaphysical consequences from quantum mechanics. Chapter 1 presents two distinctively quantum mechanical phenomena—interference and entanglement—and lays out the theory that was devised to account for them. The notable thing about quantum mechanics is that it is remarkably silent about what the basic mathematical structures of the theory represent. Chapter 2 explores the possibility that this silence is because no descriptive account of the physical reality behind quantum phenomena is possible: Quantum mechanics is just a very good predictive recipe. If this is right, of course, then quantum mechanics tells us nothing at all about ontology. But although there are theorems that preclude some ways of describing the underlying physical reality, I argue that realism about the quantum world is perfectly tenable. Chapter 3 presents three incompatible ways of understanding the reality behind the quantum phenomena—spontaneous collapse theories, hidden variable theories, and many-worlds theories—and discusses the extent to which the choice among these theories is underdetermined by the empirical data.

The next five chapters investigate the consequences of quantum mechanics for various metaphysical issues. Chapter 4 is about indeterminacy: To what extent does quantum mechanics entail that there is vagueness in the world, as opposed to in our language? Chapter 5 looks at the consequences of quantum mechanics for the kinds of causal explanation we give for physical phenomena, and in particular, at whether quantum mechanics requires action at a distance. Chapter 6 is about determinism: Does quantum mechanics entail that the physical laws are fundamentally probabilistic, and what does it tell us about the nature of probability? Chapter 7 investigates the dimensionality of the world according to quantum mechanics and evaluates arguments that the apparent three-dimensionality of the world is an illusion. And Chapter 8 examines whether quantum mechanics requires holism—that wholes have properties that cannot be reduced to the properties of their parts—and, if so, what this tells us about the ontological priority of parts and wholes and about the existence of localized individuals. The final chapter sums it all up.

1| Phenomena and Theory

ONE OF THE CENTRAL CLAIMS of this book, outlined in the Introduction, is that it is empirical quantum *phenomena* that require us to revise our classically inspired metaphysical presuppositions. To make good on that claim, I need to describe some of these phenomena, and that is the first order of business in this chapter. I then proceed to describe the theory of quantum mechanics that was developed to account for such phenomena. Typically in scientifically informed metaphysics, we look to scientific *theories* to inform our ontology. But it quickly becomes apparent why that strategy won't work in the case of quantum mechanics. First, there are two canonical formulations of the theory, so it is not clear which we should take as metaphysically privileged. Second, it is far from clear how to take either of these theories as descriptive of the world. We are left with the project of this book: If the theory of quantum mechanics by itself doesn't tell us how to conceive of the world behind the quantum phenomena, then it is up to us to construct such a conception. Fortunately, a lot of work has already been done in this direction; this is the project of *interpreting* quantum mechanics.

I do not intend here to give a historical account of the phenomena that led to the development of quantum mechanics; an excellent informal treatment can be found in Gamow (1966). The story is fascinating in its own right, but as with the development of most theories, the development of the theory and the discovery of the phenomena that require such a theory are entangled together in complicated ways. It is much cleaner, from an expository perspective, to proceed ahistorically. So in the next two sections I will present two examples of physical phenomena that most clearly demonstrate the failure of our classical intuitions and the need for a theory like quantum mechanics. It wasn't these actual phenomena that motivated the development and adoption of quantum mechanics, but they get at the heart of quantum behavior. Then in the following two sections I present the two standard mathematical formulations of quantum theory: matrix mechanics and wave mechanics. In the final section, I make the case that the theory of quantum mechanics, perhaps uniquely in the history of science, is a theory in need of interpretation.

1.1 Interference

The first phenomenon I want to discuss is interference. Interference phenomena were known long before the quantum mechanical era, notably in the behavior of light waves. Suppose you shine a light at a screen with two slits in it, and project the result on a further screen, as shown in Figure 1.1. Initially, suppose that the left-hand slit is open but the right-hand slit is blocked; then the screen is illuminated behind the open slit, as shown in the top diagram. Similarly, if the right-hand slit is open and the left-hand slit is blocked, the screen is illuminated behind the right-hand slit, as shown in the middle diagram. But if both slits are open, the illumination on the screen is not the sum of those in the top and middle diagrams; rather, it is the pattern of light and dark bands shown in the bottom diagram.

This behavior can be readily understood in terms of the mechanics of light waves, as shown in Figure 1.2. The waves spread out from the source until they encounter the first screen, where most of the wave is blocked. When one slit is open, as in the top diagram, some of the wave passes through the open slit and spreads out until it hits the second screen. This explains the illumination of the screen behind the open slit. When both slits are open, as in the bottom diagram, some of the wave passes through each slit, and the two wave components spread out and overlap before they hit the second screen. In this overlap region, interference occurs; that is, the waves from the two slits add up in some directions and cancel out in others. The solid lines represent the wave peaks, and in some directions (marked by dashed lines) you can see that the lines emanating from each slit cross, meaning that the peaks coincide. In these directions the waves from the two slits reinforce each other, resulting in a bright band on the screen. Between these directions, the peaks from one slit coincide with the troughs from the other; here the waves from the two slits cancel each other out, resulting in a dark band on the screen.

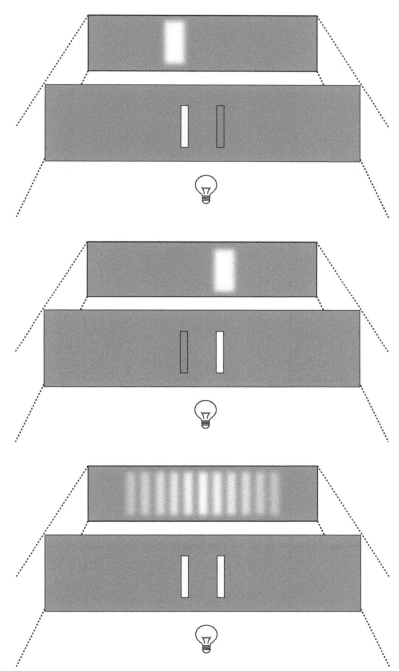

FIGURE 1.1 Two-slit interference with light.

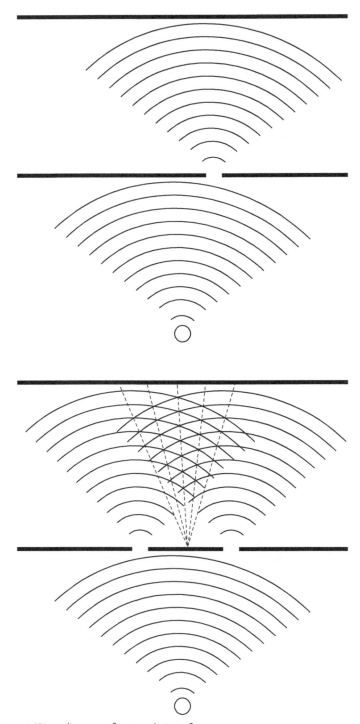

FIGURE 1.2 Wave diagrams for two-slit interference.

So far, this is just a classical wave phenomenon. A more surprising phenomenon—and a paradigmatically quantum mechanical one—is the production of an interference pattern like this using a stream of electrons instead of a beam of light. Suppose that the light source is replaced by an electron gun like that at the back of an old-fashioned television, and the display screen is replaced by a fluorescent screen like that on the front of such a television. Then you can produce interference effects exactly like those produced by light: When only one slit is open, the screen is illuminated behind the relevant slit, but when both slits are open, the screen displays a pattern of light and dark bands. This behavior is utterly inexplicable from a classical point of view. Classically, electrons are viewed as particles—essentially as like little billiard balls traveling through space. Whereas a wave can pass through both slits and then interfere, a particle, by its nature, has to pass through one slit or the other, so any explanation of interference is ruled out. That is, if the electron passes through the left-hand slit, then it can make no difference to the path of the electron whether the right-hand slit is open or not, so the pattern when both slits are open *has* to be simply the sum of the single-slit patterns.

So perhaps we were wrong about the nature of electrons: Perhaps they are really waves, not particles. This is what Louis de Broglie (1924) proposed—that every so-called particle can be treated as a wave with wavelength $\lambda = h/mv$, where m is the mass of the particle, v is its velocity, and h is a fundamental constant called Planck's constant. These wavelengths are very small; for an electron at typical lab velocities, the de Broglie wavelength is around a nanometer (a billionth of a meter), compared to several hundred nanometers for visible light. This means that to actually observe interference effects for electrons, one needs to use a slit spacing of atomic dimensions; typically, the lattice of atoms in a crystal is used in place of slits.

So far, the explanation of electron interference doesn't require any fundamental metaphysical reorientation: Electrons are reclassified as waves rather than particles, but the basic classical ontology of waves and particles remains intact. In fact, though, the situation is far more mysterious than it at first appears. Suppose we reduce the firing rate of the electron gun so that the electrons pass through the slits one at a time. Each electron produces a flash at a single precise point on the fluorescent screen, but if we look at the distribution of the flashes over time, we see that it forms an interference pattern. A typical cumulative pattern of flashes is shown in Figure 1.3. Note that the interference pattern is present even though the electrons pass through the apparatus singly; each electron interferes with *itself*, not with other electrons. But more problematically, the explanation

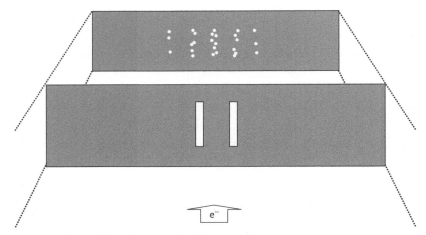

FIGURE 1.3 Two-slit interference using electrons.

for what we see seems to require that we treat the electron *both* as a wave and as a particle. The explanation of the interference pattern apparently requires that each electron behaves as a wave, passing through both slits. But the explanation of the discrete flashes apparently requires that each electron behaves as a particle, and a particle passes through one or the other slit.

In fact, exactly the same phenomenon can be observed with light. If the light intensity is reduced sufficiently, then the light arrives at the display screen in the form of flashes at precise locations, and these flashes build over time to produce an interference pattern just as in Figure 1.3. So even what we took to be an entirely classical wave phenomenon is in fact a deeply mysterious quantum phenomenon, exhibiting both wave-like and particle-like aspects. If we use the word "photon" to refer to the particle-like constituents of light, then the interference of light can be given the same paradoxical "explanation" as electron interference: Each photon behaves as a particle in that it produces a flash at a point on the screen, but as a wave passing through both slits in that the pattern of flashes exhibits interference effects. Physicists are apt to describe this situation as "wave-particle duality," but naming the mystery doesn't make it any less mysterious. The pseudo-classical "explanation" of the phenomenon is flat-out contradictory, describing the same entity simultaneously as a discrete particle and as a spread-out wave. We cannot simply combine elements of our classical metaphysical picture to form an inconsistent hybrid; if we are to solve this mystery, it seems, we need to replace our classical metaphysics with something else entirely.[1]

1.2 Entanglement

The second quantum phenomenon I want to describe is entanglement. The most straightforward way to introduce entanglement phenomena is in terms of a property of particles called *spin*. Some particles behave as if they have angular momentum like spinning balls, and some charged particles behave as if they have a magnetic field like spinning charged balls. One should probably not take this image of spinning balls literally; for one thing we have not established that quantum "particles" are little balls in any sense yet. So for the moment we should ignore the implications of the name and forgo trying to picture spin in any way at all.

A central aspect of quantum phenomena is that many physical properties are *quantized*—discrete rather than continuous. According to classical mechanics, properties like energy and momentum can take a continuous range of values. However, discoveries in the early 20th century suggested that in fact these quantities can take only certain discrete values and not any of the values in between. Indeed, it is these phenomena that provided the initial motivation for the development of quantum mechanics, as well as the name of the theory. But quantization as such is only mildly metaphysically revolutionary: If the lesson of quantum mechanics is just that energy comes in discrete chunks, then quantum metaphysics would be of only passing interest. Interference and entanglement suggest that the lessons go much deeper.

Spin is no exception to quantization: The spin of a particle can only take certain discrete values. For a charged particle like an electron, since its spin is associated with a magnetic field, these values can be measured by passing the electron through another magnetic field and observing the deflection caused by the interaction of the fields.[2] What one finds is that every electron exhibits one of two deflections: Either the electron is deflected upward by a certain amount (relative to the direction of the external magnetic field), or it is deflected downward by the same amount. That is, the spin of the electron can take only two possible values relative to the external field, called (for obvious reasons) "spin-up" and "spin-down." Particles of this kind—those that can take only two spin values—are called spin-1/2 particles.[3]

More surprisingly from a classical perspective, the spin of the electron can take only these two values relative to *any* direction of external field. Suppose the external field is initially aligned vertically, and that a given electron is deflected upward relative to this field. Thinking classically, one might think that the magnetic field of the electron must be aligned vertically, and hence that it must be spinning around an axis that is aligned vertically.

Such thinking implies that if one then passes the electron through an external field aligned horizontally, it should be deflected less than in the vertical case (and not at all if the field is completely homogeneous), since the applied field is now perpendicular to the field of the electron. But in fact the electron is deflected by exactly the same amount as before, either upward or downward relative to the new field. That is, the electron is either spin-up or spin-down relative to any direction one might pick. This provides yet another reason not to think of quantum spin in terms of particles literally spinning.

So the quantization of spin is mysterious in itself from a classical point of view. But the phenomenon of entanglement compounds the mystery. The basic idea is simple enough. Because spin is a conserved quantity, if a spinless particle decays into two spin-1/2 particles, the spins of the resulting particles must be equal and opposite. This can be verified experimentally. For pairs of spin-1/2 particles produced by the decay of a spinless particle, if one measures their spins in the same direction, one always finds that one of them is spin-up and the other is spin-down. Such pairs of particles are called "entangled"; their spins are correlated rather than independent.

Now there is nothing mysterious in the correlation per se. It is perfectly explicable in terms of the origin of the pair of particles. What is mysterious is how this correlation is instantiated in the properties of the individual particles. This becomes apparent if we measure the spins of the two particles in different directions. We know that if we measure the spins of the particles in the same direction, the results always disagree, and by the same token, if we measure the spins in opposite directions the results always agree. But what about intermediate directions? This is easy to check, and the result is that for intermediate angles the results agree some of the time and disagree some of the time. More precisely, if the angle between the two measuring devices (the two external fields) is θ, the probability of agreement is given by $\sin^2(\theta/2)$. So when the two measuring devices are aligned 120° apart, for example, the spins of the two particles agree $\sin^2 60° = 3/4$ of the time.

Still, this does not seem so hard to explain, at least at first glance. After the two particles have gone their separate ways, they presumably have their individual spin properties, and these properties are correlated. What we need for a full explanation of the experimental results is some recipe for assigning spins to particles so as to reproduce the observed distribution of measurement results. Suppose that the first particle is spin-up along the vertical axis; then the observed distribution requires that the second particle is spin-down along the vertical axis, and also has a 3/4 chance of being spin-up along a direction 120° from vertical. But in 1964, Bell proved that, subject to some physically and metaphysically plausible assumptions, no such recipe exists; it is impossible to consistently assign spin properties to the individual particles that reproduce the observed distribution of measurements results (Bell 1964). Bell's theorem and its consequences will be explained in Chapter 2, but the immediate import of the theorem is that unless we can find a way to reject one of Bell's plausible assumptions, we have to deny that the correlations manifested by the pair of entangled particles can be explained in terms of the properties of the individual particles. Because we have no way of explaining a correlation *other* than by appealing to the properties of the individual correlated entities, entanglement presents us with another mystery.

So far I have said nothing about the *theory* of quantum mechanics. This is deliberate: My goal in this section was to show that the metaphysically problematic nature of quantum mechanics is not just a matter of interpreting an obscure theory, but is a problem in the empirical phenomena themselves—in the behavior of objects in the world. The world does not conform to our classically trained intuitions. So regimenting these phenomena by subsuming them under a theory is not going to solve our metaphysical difficulties. Still, the regimentation provided by the theory of quantum mechanics is a very important step in understanding what our metaphysical options are, since the rules embodied by the theory are (as far as we know) exceptionless within its domain of application. If nothing else, the theory provides a useful summary of the metaphysically problematic phenomena we need to address. So let us see what it says.

Historically, *two* theories of quantum phenomena were developed at the same time—the matrix mechanics of Heisenberg (1925) and Born and Jordan (1925), and the wave mechanics of Erwin Schrödinger (1926a). The two theories were quickly shown to be mathematically equivalent, but because they represent the world in prima facie distinct ways, it is worth introducing them separately. Furthermore, as we shall see, the two quantum phenomena just described naturally line up with the two theories: Interference is more naturally represented using wave mechanics, whereas entanglement of spins is more naturally represented using matrix mechanics. I will start with the latter.

First, a small terminological matter. Physicists use the word "state" to refer both to the physical state of a system and to the mathematical representation of that physical state. This is harmless in physics, but confusing in philosophy, where one is often trying to discuss the relationship between the representation and the physical world. So where ambiguity threatens, I will use the phrase "physical state" to refer to the state of the physical world and the phrase "quantum state" to refer to the mathematical representation in the theory.

1.3 Matrix Mechanics

The theory of quantum mechanics, however formulated, has three elements: the quantum state that represents the physical system we are interested in, the dynamical law by which the quantum state changes over time, and the measurement postulate that relates the quantum state to the results of measurements. In matrix mechanics, the physical state of a system is represented using a vector—that is, as a list of numbers.[4] The number of entries in the list depends on the number of basic distinct physical states of the system. If there are two such states, the list has two numbers in it, if there are three, the list has three numbers in it, and so on.

Thinking pictorially, one can regard the list of numbers as specifying the coordinates of the endpoint of an arrow that starts at the origin. The arrows representing the basic distinct states of the system are mutually orthogonal—they lie at right angles to each other—and of length 1. Hence, the simplest way of representing the two basic states of a two-state system is (1,0) and (0,1). So for example, since an electron has two distinct basic spin states relative to the z-axis, spin-up and spin-down, these two vectors can be used to represent the basic states. That is, we can write $|\uparrow_z\rangle = (1,0)$ and $|\downarrow_z\rangle = (0,1)$, where the asymmetric brackets $|\rangle$ are Dirac's notation for a quantum state. Here $|\uparrow_z\rangle$ is shorthand for "the quantum state that represents the electron as spin-up relative to the z-axis" and $|\downarrow_z\rangle$ is shorthand for "the quantum state that represents the electron as spin-down relative to the z-axis."

In textbooks, the vectors that represent the basic distinct physical states of a system are called the *eigenstates* (or *eigenvectors*) for the physical property concerned, and given a rigorous mathematical definition. A physical property like z-spin is represented by an operator on the space of vectors—a mathematical object that transforms the vectors in the space into other vectors in the space. The vectors whose direction is unchanged by this operation are the eigenstates of the operator. So if $\widehat{S}_z$ is the operator representing z-spin, then the eigenstates of $\widehat{S}_z$ are the vectors satisfying $\widehat{S}_z|\psi\rangle = c|\psi\rangle$, where c is a real number—the vectors such that the effect of applying the operator to them is to multiply their length by a factor of c while leaving their direction unchanged. The factor c is called the eigenvalue for the eigenstate, and represents the spin of the electron when its state is

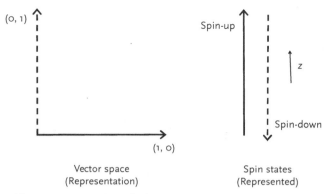

FIGURE 1.4 Vector representation of spin states.

the eigenstate. The operator $\widehat{S}_z$ has two eigenstates, (1,0) and (0,1), with eigenvalues $+\hbar/2$ (spin-up) and $-\hbar/2$ (spin-down), respectively, where $\hbar$ is Planck's constant h divided by 2π.

Figure 1.4 contains (on the left) a diagram of the space containing these two vectors: The spin-up vector (1,0) is a unit-length vector along the horizontal axis, and the spin-down vector (0,1) is a unit-length vector along the vertical axis. Such diagrams are apt to be confusing: The two eigenstates lie at right angles to each other in the vector space, but the basic spin states they represent are up and down relative to *the same* spatial direction, as shown on the right. If one thinks of spin-up and spin-down as arrows pointing up and down along the z-axis, the two spins are 180° apart, whereas the two vectors are 90° apart. So it is important to distinguish carefully between the directions of the vectors in the vector space and the directions of the spins in real space: They are not the same.

Note also that even if we restrict ourselves to unit-length vectors, the vector space shown in Figure 1.4 contains plenty of other vectors—all those whose endpoint lies on a unit-radius circle centered on the origin, like those shown on the left in Figure 1.5. That is, any vector of the form (a,b) satisfying $|a|^2 + |b|^2 = 1$ (so that the length is 1) is a vector in the space.[5] All these vectors are possible quantum states of the system, and matrix mechanics makes essential use of the full range of such states in its explanations. But by hypothesis, this system has two basic distinct states; how then can it also have a continuum of possible states? What do the vectors other than the two eigenstates represent? This is one of the central questions in the interpretation of quantum mechanics, and one we will be grappling with in some form or other for much of the book.

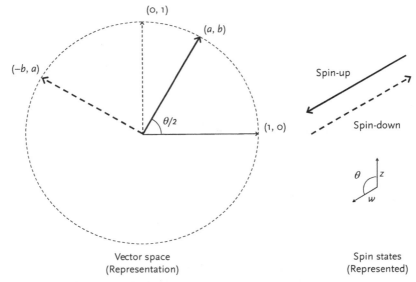

FIGURE 1.5 Other vectors in the space.

Mathematically, the relationship between the eigenstates and the rest of the vectors in the space is straightforward: The eigenstates form a *basis*, meaning that every vector in the space can be expressed as a weighted sum of the eigenstates. That is, any vector (a,b) in the space can be rewritten as the sum $a(1,0) + b(0,1)$. It is common to call the vector (a,b) a *superposition* of the basic states $(1,0)$ and $(0,1)$, because the former can be obtained by superposing (adding together) the latter (suitably weighted). But note that there is nothing special about the vectors $(1,0)$ and $(0,1)$ in this respect: any pair of orthogonal vectors in the space can act as a basis. For example, the vectors (a,b) and $(-b,a)$ shown in Figure 1.5 form a basis, and the vector $(1,0)$ can be represented as a weighted sum of (a,b) and $(-b,a)$.[6] So mathematically speaking, no vector is intrinsically a superposition.

This mathematical "democracy of bases" has a physical analog in terms of operators: There is an operator that has (a,b) and $(-b,a)$ as eigenstates, corresponding to a different choice of basic physical states we could have made. After all, there is nothing physically privileged about the z-axis. Suppose that instead of spin-up and spin-down along the z-axis, we choose spin-up and spin-down along direction w as our basic states, where w makes an angle θ to z, as shown on the right in Figure 1.5. The physical property of w-spin (spin relative to the w-axis) is represented by an operator that has vectors (a,b) and $(-b,a)$ as eigenstates, where $\cos(\theta/2) = a$ and $\sin(\theta/2) = b$.

So, for example, if w makes an angle of 120° with z, then $a = \cos 60° = 1/2$ and $b = \sin 60° = \sqrt{3}/2$. Hence, the vectors $|\uparrow_w\rangle = (1/2, \sqrt{3}/2)$ and $|\downarrow_w\rangle = (-\sqrt{3}/2, 1/2)$ represent states in which the electron is spin-up and spin-down relative to w, respectively: They are the eigenstates of the operator $\widehat{S}_w$ representing spin along w.

So we have a partial interpretation of the continuum of vectors in the vector space. Vectors $(1,0)$ and $(0,1)$ represent states in which the particle is spin-up and spin-down in the z direction, $(1/2, \sqrt{3}/2)$ and $(-\sqrt{3}/2, 1/2)$ represent states in which the particle is spin-up and spin-down in the w direction, and similarly for all the other vectors in the space. But this is only a partial interpretation: It doesn't answer the question of what the spin of the particle is in the z direction when its state is $(1/2, \sqrt{3}/2)$, for example. This state isn't an eigenstate of spin in the z direction: It lies between the eigenstates. But it can't represent an intermediate z-spin value between $+\hbar/2$ (spin-up) and $-\hbar/2$ (spin-down) because spin is quantized: Only those two values are allowed. So what does it represent? Note that the problem here is *not* that some states can be readily interpreted in terms of the properties of the system and some cannot. Every state is an eigenstate of some operator, and hence can be regarded as representing the system as possessing the corresponding property. But every state is also a *noneigenstate* of other operators, and it is for these properties that we don't know what to say. The state $(1,0)$ is no less problematic than the state $(1/2, \sqrt{3}/2)$: The former state is an eigenstate of z-spin, but it is not an eigenstate of w-spin, so although we know what z-spin property it represents the system as having, we don't know what it says about w-spin.

The theory of quantum mechanics notably doesn't address these issues; it quietly changes the subject. Rather than answering the question "What do the noneigenstate vectors for an operator represent?" it answers the question "What happens when we measure a system whose state is not an eigenstate of the operator corresponding to the measured property?" The representation itself is silent on this question, so a separate postulate is added to the theory specifically to answer it. The *measurement postulate* says that if the spin of an electron in state (a,b) is measured in the z direction, one obtains result "up" with probability $|a|^2$ and "down" with probability $|b|^2$. Pictorially, the probability of obtaining spin-up can be obtained by *projecting* the state vector (a,b) onto the vector $(1,0)$ corresponding to z-spin-up, and squaring the result; this is called the Born rule (Born, 1926). Similarly, the probability of obtaining spin-down can be obtained by projecting (a,b) onto $(0,1)$ and squaring; both projections are shown in Figure 1.6. Because projection is a symmetric operation, the result of measuring the spin of an

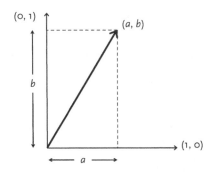

FIGURE 1.6 Projections of (a,b) onto $(1,0)$ and $(0,1)$.

electron in state $(1,0)$ in the direction corresponding to (a,b) is the same as the result of measuring the spin of an electron in state (a,b) in the direction corresponding to $(1,0)$: Projecting $(1,0)$ onto (a,b) and squaring also yields a probability of $|a|^2$ for "up." This symmetry is made evident by a simple trick for calculating the projection of one vector onto another: Multiply the corresponding entries in each vector together, and add the result. That is, the projection of (a_1, b_1) onto (a_2, b_2) is $a_1 a_2 + b_1 b_2$, and the probability of getting spin-up for the corresponding measurement (of an electron in state (a_1, b_1) in the direction defined by (a_2, b_2), or vice versa) is the square of this projection.[7]

It is worth noting that the projection involved here is usually conceived as a physical process, not a mere calculational exercise. The reason is that successive measurements always yield the same result: If you measure the z-spin of an electron in state (a,b) and obtain spin-up, and then you measure its z-spin a second time, you are guaranteed to obtain spin-up again. That is, the probability of obtaining spin-up on the second measurement is 1, so the results of the second measurement aren't obtained by applying the Born rule to the original state (a,b). Instead, it is typically supposed that the first measurement *projects* or *collapses* the state onto the eigenstate corresponding to the outcome you obtain, in this case $(1,0)$. Then the second measurement is a z-spin measurement on the z-spin up eigenstate, which by the Born rule yields "up" with a probability of 1 as required. For this reason, the measurement postulate is sometimes known as the projection postulate or the collapse postulate. This dynamical feature of the measurement postulate will be explored in detail in Chapter 3.

The final thing we have to add to obtain a mechanical theory is a dynamical law—a law that specifies how the state of a system changes over time. In quantum mechanics, this law is the Schrödinger equation:

$$i\hbar\frac{\partial}{\partial t}|\psi\rangle = \hat{H}|\psi\rangle \tag{1.1}$$

Here $|\psi\rangle$ is the quantum state of the system under consideration, and $\partial/\partial t|\psi\rangle$ is its time derivative—the rate of change of $|\psi\rangle$. The factor $i\hbar$ is a constant, and $\hat{H}$ is the operator corresponding to the physical property of energy. The form of this operator depends on the nature of the system we are modeling. An important property of this dynamical law, one that we will refer to frequently over the subsequent chapters, is that it is *linear*. This means that if an initial state $|\psi_1\rangle$ evolves according to the Schrödinger equation to the final state $|\psi_1'\rangle$, and an initial state $|\psi_2\rangle$ evolves to the final state $|\psi_2'\rangle$, then any linear superposition $a|\psi_1\rangle + b|\psi_2\rangle$ of the initial states evolves to the analogous linear superposition $a|\psi_1'\rangle + b|\psi_2'\rangle$ of the final states.

So we now have the outlines of the theory of quantum mechanics: Every state of a system is represented by a vector in a vector space, the vector changes over time according to the Schrödinger equation, and the results of measurements on the system are given by the Born rule. Let us apply this theory to the example of entangled electrons we looked at in the previous section. Because entanglement involves the spin states of a pair of particles at a single time, we do not have to consider changes in the quantum state over time, so we can ignore the dynamical law for the moment.

Entanglement, recall, is a feature exhibited by a pair of particles, and a pair of spin-1/2 particles has four distinct basic spin states relative to the z-axis. That is, both electrons can be z-spin-up, or both can be z-spin-down, or the first can be z-spin-up and the second z-spin-down, or the first can be z-spin-down and the second z-spin-up. So to represent this system using matrix mechanics, we need a four-dimensional vector space spanned by four mutually orthogonal vectors:

$$|\uparrow_z\rangle_1|\uparrow_z\rangle_2 = (1,0,0,0) \tag{1.2}$$

$$|\uparrow_z\rangle_1|\downarrow_z\rangle_2 = (0,1,0,0)$$

$$|\downarrow_z\rangle_1|\uparrow_z\rangle_2 = (0,0,1,0)$$

$$|\downarrow_z\rangle_1|\downarrow_z\rangle_2 = (0,0,0,1)$$

where $|\uparrow_z\rangle_1|\uparrow_z\rangle_2$ is shorthand for "the quantum state in which the first electron is z-spin-up and the second electron is z-spin-up," and similarly for the others.

Note that the second and third vectors in (1.2) represent the two electrons as having opposite z-spins, as entangled electrons do. However, the entangled state cannot simply be $|\uparrow_z\rangle_1|\downarrow_z\rangle_2$, since it is not always the case that the first electron is spin-up and the second is spin-down; half the time it is the other way around. By the same token, the entangled state cannot be $|\downarrow_z\rangle_1 |\uparrow_z\rangle_2$. Instead, we represent the entangled state $|S\rangle$ of the pair of electrons by combining these two basic vectors as follows:

$$|S\rangle = \frac{1}{\sqrt{2}}(|\uparrow_z\rangle_1 |\downarrow_z\rangle_2 - |\downarrow_z\rangle_1 |\uparrow_z\rangle_2) \qquad (1.3)$$
$$= (0, 1/\sqrt{2}, -1/\sqrt{2}, 0).$$

Now we can apply the Born rule to see what happens when two electrons in this entangled state have their z-spins measured. The probability of getting result "up" for both electrons is obtained by projecting the entangled state $|S\rangle$ onto the basic state $|\uparrow_z\rangle_1 |\uparrow_z\rangle_2$ and squaring the result. To calculate the projection, we can use the trick mentioned earlier; multiply the corresponding entries in each vector together and add. This yields a probability of zero. The probability for getting result "down" for both electrons can be calculated in the same way, again yielding zero. Similarly, the probability of getting "up" for the first and "down" for the second is 1/2, as is the probability of getting "down" for the first and "up" for the second. That is, when the electrons' spins are measured in the same direction, the results never agree.

But what if we measure the spins of the two electrons along different directions—say, we measure the first electron along z and the second along w (where w, recall, makes an angle of 120° with z)? To calculate the relevant projections, we need to construct the vectors corresponding to the first electron being z-spin-up or z-spin-down and the second being w-spin-up or w-spin-down. We can do this by noting that the first two slots in each vector correspond to the first electron being z-spin-up, so the vectors in which the first electron is w-spin-up can be constructed by plugging the w-spin-up eigenstate $(1/2, \sqrt{3}/2)$ and the w-spin-down eigenstate $(\sqrt{3}/2, -1/2)$ into these two slots. Similarly, the last two slots correspond to the first electron being z-spin-down, so the remaining two vectors can be constructed by plugging $(1/2, \sqrt{3}/2)$ and $(\sqrt{3}/2, -1/2)$ into these slots. The result is the following set of vectors:

$$|\uparrow_z\rangle_1 |\uparrow_w\rangle_2 = (1/2, \sqrt{3}/2, 0, 0) \qquad (1.4)$$
$$|\uparrow_z\rangle_1 |\downarrow_w\rangle_2 = (\sqrt{3}/2, -1/2, 0, 0)$$
$$|\downarrow_z\rangle_1 |\uparrow_w\rangle_2 = (0, 0, 1/2, \sqrt{3}/2)$$
$$|\downarrow_z\rangle_1 |\downarrow_w\rangle_2 = (0, 0, \sqrt{3}/2, -1/2).$$

We can then repeat the Born rule calculation for this new set of vectors, to see what results we should expect if we measure the first electron along z and the second along w. Projecting $|S\rangle$ onto $|\uparrow_z\rangle_1|\uparrow_w\rangle_2$ and squaring yields a probability of 3/8, and projecting $|S\rangle$ onto $|\downarrow_z\rangle_1|\downarrow_w\rangle_2$ and squaring again yields a probability of 3/8. That is, if the electrons have their spins measured along two axes 120° apart, the results agree 3/4 of the time. This, of course, is precisely what we observe: This theory successfully reproduces the mysterious correlations exhibited by entangled particles.

1.4 Wave Mechanics

The matrix mechanics outlined in the previous section is ideal for modeling properties like spin that take just a few discrete values. But what if we want to represent a property like position? The position of an electron can presumably take a continuum of possible values.[8] How can we represent a continuous quantity using the aforementioned theory?

One way to approach things is to start with a discrete approximation of the continuous quantity, and then derive the continuous representation as a limiting case. Suppose we want to model the position of an electron moving along one dimension within a particular region of space. Initially, let us divide up the region into a number of smaller regions such that the electron occupies one of them: these are the basic physical states of our system. Then we can straightforwardly apply matrix mechanics: If we have divided the region into five basic locations, we need a five-dimensional vector space spanned by five mutually orthogonal vectors. So (1,0,0,0,0) represents a physical state in which the electron is in the first region, (0,1,0,0,0) represents a state in which the electron is in the second region, and so on. These vectors are the eigenstates of an operator corresponding to this coarse-grained location property for the electron. A general vector in this space can be written (a,b,c,d,e), where a,b,c,d, and e are (complex) numbers such that $|a|^2 + |b|^2 + |c|^2 + |d|^2 + |e|^2 = 1$. For reasons that will become obvious, we will call a,b,c, and so on the *amplitudes* associated with their respective locations.

We can picture these quantum states by means of bar charts like the one in the top left of Figure 1.7: Each bar represents a spatial location, where the area of the leftmost bar is a, the area of the next one is b, and so on. So the eigenstates for the location of the electron are the states in which only one bar has nonzero area, and noneigenstates like the one depicted are

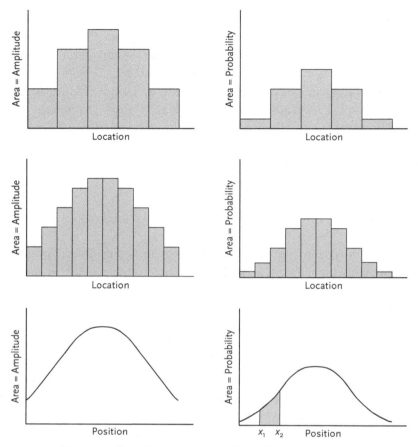

FIGURE 1.7 Discrete and continuous representations of position.

such that several or all of the bars have nonzero area. (Again, don't ask what these noneigenstates represent!) The probability of finding the electron in each location is given by the square of the area of the corresponding bar in the top-left graph, and these probabilities are graphed explicitly in the bar chart in the top right of Figure 1.7: Here the area of the first bar is $|a|^2$, the area of the second bar is $|b|^2$, and so on, so that the total area under the bar chart is 1.

This pictorial representation makes it easy to imagine adopting a more fine-grained representation of the position of the electron. We could divide the region into twice as many locations of half the size as before, so that the bar charts in the middle row of Figure 1.7 have twice as many bars of half the width as those in the top row—so again the total area under the squared-amplitude bar chart on the right is 1. As the number of locations gets larger and the width of the bars gets smaller, the bar charts approach the smooth curves shown in the bottom row.[9]

The curve in the bottom left diagram is called the *wave function* of the electron, usually written $\psi(x,t)$ since it is a function of the position coordinate x and also varies over time. It is the standard representation of the (one-dimensional) position state of an object in Schrödinger's wave mechanics. The curve in the bottom right diagram is the squared-amplitude function $|\psi(x,t)|^2$; it is the continuous limit of the bar chart above it, and the square of the wave function (technically the absolute square, since the wave function can in general take complex values). As before, the area under the squared-amplitude function is related to probability. In the continuous case, the probability of finding the electron in some spatial region bounded by x_1 and x_2 at time t is given by the area under the curve between these points—or in other words, by the integral $\int_{x_2}^{x_1} |\psi(x,t)|^2 \, dx$. This is the form that the Born rule takes in wave mechanics. The total area under the squared-amplitude function is 1.

It is easy to generalize this approach to three dimensions—just harder to draw. One can imagine dividing a three-dimensional region of space into a number of small cubes, and having these serve as the coarse-grained locations in the analysis. As the size of these cubes gets smaller, the three-dimensional "bar chart" approximates a smooth function of the three spatial dimensions, which we can write as the three-dimensional wave function $\psi(x,y,z,t)$. This is the standard wave-mechanical representation of the position state of an electron in three dimensions. The probability of finding the electron in some three-dimensional region R at time t is given by integrating the corresponding squared-amplitude function over R, that is, by $\int_R |\psi(x,y,z,t)|^2 \, dx\,dy\,dz$. But for the moment I will stick to one spatial dimension.

So we can use this method to move from a discrete vector representation to a continuous wave function representation. It is worth noting, though, that there is nothing in principle to prevent us from using the vector representation for continuous quantities like position: We just need to divide space into discrete ranges, which can be made as small as we like.[10] Conversely, there is nothing to prevent us from using the wave function representation for discrete quantities like spin, although the functions in this case will be discrete rather than continuous. This is the sense in which the two theories are equivalent: Each can cover the full range of cases, but it is usually easier to apply wave mechanics to continuous quantities and matrix mechanics to discrete ones.

Note also what happens to the basic position states when we move from a discrete to a continuous representation. In the discrete case, the basic physical states are coarse-grained location ranges for the electron, represented by bar charts with only one nonzero bar. In the continuous case, the basic physical states are presumably precise spatial positions of the electron. But how are these basic states represented in the theory? As the bar chart gets more and more fine-grained, the basic physical states are represented using narrower and narrower bars, and since the area of the single bar has to be 1, the bar has to get taller. In the continuous limit, the basic states would have to be represented by squared-amplitude functions whose value is nonzero only at a single point—infinitely tall, thin functions—but which still satisfy the requirement that the area enclosed by this thin spike is 1. But there are no such functions: It is a mathematical fact that no function of zero width can enclose a nonzero area.[11] What this means is that in continuous wave mechanics, unlike in discrete matrix mechanics, the basic states are not represented in the theory. This makes our interpretive problem more severe. In matrix mechanics, there is at least a correspondence between some states and some properties—z-spin eigenstates and z-spin properties, for example. In wave mechanics, there is no obvious way to associate *any* actual quantum state with any position property, because there are no quantum states corresponding to precise positions.[12]

Wave mechanics ducks this issue in the same way as matrix mechanics—by stressing prediction rather than representation. Ask not "What does this quantum state represent?"; ask "What can I expect when I measure it?" The answer to the latter question is given by the Born rule, now formulated as an integral. Note that we cannot find the probability that the electron has a particular precise position by performing the wave function analog of projecting the quantum state onto the eigenstate representing this position, since there are no such states in our representation. But there are reasons not to worry about this: One cannot measure the position of anything with perfect precision, and even if we could, the probability that an electron has some *precise* position is presumably zero. Still, this only addresses the epistemic question; it doesn't even pretend to address the question of whether electrons have precise positions.

This takes care of the representation of physical states and the measurement postulate in wave mechanics; all that remains is the dynamical law. This is the same Schrödinger equation, now expressed in terms of continuous functions rather than vectors:

$$i\hbar \frac{\partial}{\partial t} \psi(x,t) = \hat{H} \psi(x,t) \tag{1.5}$$

The energy operator in this context is a differential operator: It turns one function of x into another function of x, and again depends on the system under consideration. But a particular case worth is worth mentioning, namely a free particle—a particle not subject to external forces. In that case, the energy of the system is just the kinetic energy of the particle, and the operator takes the form $-(\hbar^2/2m)\partial^2/\partial x^2$, where $\hbar^2$ is a constant, m is the mass of the particle, and $\partial^2/\partial x^2$ is the second derivative with respect to position. So in that case the Schrödinger equation becomes the differential equation

$$i\hbar \frac{\partial}{\partial t}\psi(x,t) = -\frac{\hbar^2}{2m}\frac{\partial^2}{\partial x^2}\psi(x,t). \tag{1.6}$$

The reason this case is worth mentioning is that this is a familiar kind of equation. It is a wave equation; that is, its solutions $\psi(x,t)$ have the form of waves traveling through space. This explains the terminology in which the theory is written: *wave mechanics* and *wave function*, and also *amplitude*, since the size of ψ for a free particle is the amplitude of a wave.

What happens when we apply this theory to the example of two-slit interference? The example concerns the position of a single free electron over time, so we solve the free particle Schrödinger equation (1.6) for an electron that is initially at the source. The solutions are waves spreading out from the source, like three-dimensional ripples on a pond. Most of the wavefront hits the first screen, but some passes through whichever slits are open. An open slit acts like a new source of waves; waves spread out from the open slit. If one slit is open, this wave spreads out unimpeded until it hits the display screen, just as in the top diagram of Figure 1.2. Hence, the amplitude of the waves hitting the display screen is highest behind the open slit. If both slits are open, then the waves emanating from each slit interfere with each other: They add up in some directions and cancel out in others, just as in the bottom diagram of Figure 1.2. So in this case the wave amplitude at the display screen exhibits an interference pattern of high-amplitude bands and low-amplitude bands.

So we can apply wave mechanics to determine the quantum state of the electron at the display screen. But recall that we have as yet no physical interpretation of the quantum state; all we have is a rule that tells us what to expect when we measure the position of the electron. Accordingly, let us treat the display screen as doing just that—as measuring the position of the electron and producing a flash at a position corresponding to the result. Then we can use the Born rule to give us the probability that the electron will be found in any particular region of the screen: the probability distribution over the screen is just the square of the wave function amplitude. So when one slit is open, the electron is most likely to be found behind the open slit (Figure 1.1, top and middle), and when both slits are open, the probability distribution looks like an interference pattern (Figure 1.1, bottom). In particular, if we allow many electrons to pass through the apparatus one by one, we should expect the pattern of flashes to build up over time to produce the interference pattern (Figure 1.3). This is exactly what we actually observe.

1.5 Interpretation

We have seen how two versions of the theory of quantum mechanics can be applied to reproduce the phenomena we initially found so puzzling: interference and entanglement. But does the theory thereby *explain* those phenomena? Well, it certainly provides some of the superficial trappings of a physical explanation. The possible physical states of the systems are represented in a mathematical space, either a vector space or a space of functions, and the particular state we are interested in is represented by a particular element in that space. In the case of interference, a dynamical law dictates how the mathematical representation evolves over time, and in both cases a measurement rule connects the final representation with the outcomes we actually observe.[13]

But if quantum mechanics constitutes an explanation of interference phenomena and entanglement phenomena, you would expect it to thereby provide some kind of solution to the mystery of how such phenomena could possibly come about. Interference can't be explained via waves and it can't be explained via particles—so what kind of physical entity does quantum mechanics postulate behind the phenomena? Entanglement can't be explained via spin properties of individual electrons—so what kind of property does quantum mechanics postulate behind the observed correlations?

Unfortunately, the theory we have just invoked is entirely silent on these matters. Recall the question that is dodged by the theory: What physical states do the noneigenstates of a given operator represent? In the case of entanglement, we have no physical interpretation of the entangled state $|S\rangle$ in terms of the properties of the electrons, individually or collectively. Similarly, in the case of interference, no physical interpretation is offered of the wave functions involved in the explanation, either as waves or as particles or as something else entirely. Without a physical interpretation of the quantum state that is supposedly doing the explaining, it is not clear that we have given an explanation in any sense worthy of the term. You might reasonably suspect that we have simply given a general mathematical redescription of the observed phenomena.

This makes quantum mechanics unusual, perhaps unique, in the history of science. It is a theory in which we have no idea what we are talking about, because we have no idea what (if anything) the basic mathematical structures of the theory represent. Often in science, the interpretation of the mathematical elements of a theory is obvious. For example, in population ecology, the variables representing the number of individuals, the number of births, and the number of deaths in a population of organisms hardly need further explication. In cases where it is less obvious, presentations of the theory typically include an explicit discussion of the features of the world that the mathematical structures of the theory are supposed to represent. Newton, for example, in presenting his theory of mechanics in the *Principia*, takes great pains to explain the meanings of the terms involved, especially those like *space*, *time*, and *motion*, which differ from their ordinary meaning (Huggett, 1999, 118).

The interpretation of the mathematical elements of quantum mechanics in either of its formulations is certainly not obvious. And neither do the founders of quantum mechanics address what these elements represent in their initial presentations of the theory. The closest we get is Born's explication of the rule connecting such states to the probabilities of observing various outcomes on measurement, but this by itself doesn't tell us what these states are. Some of the founders of quantum mechanics, notably Einstein and Schrödinger, were troubled by this.

The lack of an interpretation may be troubling, but the theory of quantum mechanics is still a remarkable achievement, since it unifies so many diverse phenomena using such a concise and elegant mathematical theory. Taken as an instrumental recipe, it is practically perfect in every way: It only fails to make the correct prediction where a nonrelativistic theory would be expected to fail, namely where speeds are a significant proportion of the speed of light. But the point I wish to stress in this chapter is that even if we view the theory of quantum mechanics simply as a useful summary of empirical phenomena, it still challenges our classically tutored metaphysical intuitions. There is no obvious way to explain those phenomena in terms of the kinds of objects and properties we typically take to populate the world. Indeed, the challenge to ordinary metaphysics inherent in quantum phenomena is really the same difficulty as our inability to interpret the theory. The theory of quantum mechanics is silent on the physical interpretation of states precisely because no obvious interpretation can be given in terms of our ordinary metaphysical categories. And conversely, quantum phenomena are not explained by the theory of quantum mechanics precisely because we don't know how to take the theory as descriptive of the world.

What should we do in the face of these problems? There are several possible approaches. One approach is to try to revise our account of physical ontology in light of quantum mechanical phenomena. The hope here is that by doing so, we can find a way to take the theory of quantum mechanics (in at least one of the forms outlined earlier) as descriptive of this ontology, and thereby make the phenomena of quantum mechanics less puzzling. A second approach is to reject the claim that quantum theory (in either form) is a fully adequate theory as it stands. Perhaps it provides a *partial* description of physical systems and their behavior, and needs to be supplemented in some way with further descriptive machinery. Perhaps it shouldn't be taken as describing the world at all—perhaps it merely describes our *knowledge*—and we need to construct from scratch an alternative account of quantum phenomena that *can* be taken as descriptive of the world. These approaches, or their combination, constitute the project of *interpreting* quantum mechanics.[14]

Exploring this interpretive project takes up most of the rest of the book. But before we embark on this project, we need first to set aside a third response to the problems of quantum mechanics, which is to take them as an indication that we should give up on taking our fundamental physical theories as descriptive—that we should stop seeking explanations of quantum phenomena in terms of an underlying physical reality. That is, in order to get to the project of how to take a realist attitude toward quantum theory, we must first deal with arguments that the lesson of quantum theory is that realism is dead. This is the topic of the following chapter.

2| Realism

As we saw in the previous chapter, the distinctive phenomena of quantum mechanics—interference and entanglement—are deeply puzzling. It is hard to see how they could be given a satisfying physical explanation. The theory of quantum mechanics is of no help here because it is silent precisely where explanation usually arises: It provides no physical interpretation of the quantum states involved in modeling the quantum phenomena. So the big question is this: How could the world be such that the theory of quantum mechanics is true of it? This is the question we will be addressing in various ways over the course of the following chapters.

But before we get to this, there is a doubt about the whole project that needs to be considered. From the early days of quantum mechanics through to the present, commentators have repeatedly argued that the lesson of quantum mechanics is that physical explanation has limits. That is, the fact that quantum theory fails to explain quantum phenomena in the way we expect does not reveal some fault with quantum theory, but rather a fault in our expectations. We expect physical theories to explain by describing the underlying physical ontology and its behavior, but this model of the role of theories cannot be extended into the microphysical realm. Rather, the only proper job of quantum mechanics is as a predictive recipe—a role that it fulfills perfectly.

If this is right, then the project of this book is misconceived: There is no question of coming up with an adequate physical ontology of the quantum world. Pretheoretically, at least, we think of scientific theories as true descriptions of the world. In other words, pretheoretically we are all scientific realists, at least in van Fraassen's minimal sense that "science aims to give us, in its theories, a literally true story of what the world is like" (1980, 8). But it is not uncommon to think that quantum mechanics exposes the limits of scientific realism: Before we knew about quantum phenomena, it was quite reasonable to regard all scientific theories as descriptive, but now we must recognize that at least in the realm of microphysics, description is out of the question. It is not just that quantum mechanics, as it stands, does not describe the world; there is no description to be had. If that is so, then this book can end now: There are no metaphysical consequences of quantum mechanics.

In this chapter I lay out the case against scientific realism based on quantum mechanics. I argue that it is not conclusive, and that realism about the quantum world is very much a live option. But nevertheless the debate over realism in quantum mechanics is fascinating in its own right because of how different it is from the usual realism/antirealism debates. Typically, arguments against scientific realism are of a very general nature, based on the possibility of constructing empirically adequate alternatives to any theory whatsoever (van Fraassen, 1980), or on the poor historical track record of our theories in general (Laudan, 1981). The argument considered in this chapter doesn't concern science in general, but quantum mechanics in particular: the quantum phenomena themselves, it is claimed, preclude a descriptive account of the underlying ontology.

Note that the antirealist claim is not merely that quantum theory as it stands does not describe the world, but that it *cannot*—that no variant could do better. The failure of standard quantum theory to describe the world is entirely consistent with scientific realism. Our difficulties in interpreting quantum theory as descriptive might just show that quantum theory has faults as a theory; it doesn't threaten the thesis that the *goal* of science is literally true theories. In the next section I rehearse a classic argument to try to get more precise about the sense in which quantum mechanics fails as a description. The authors of that argument are quite explicit that its failure doesn't undermine realism, but instead calls out for a realist solution in the form of a *better* description of the quantum microworld. However, in the subsequent section I present the standard arguments—the so-called no-go theorems—that purport to show that no better description of the quantum microworld is possible. It is these results that are often taken as counting against realism. Finally, I argue that while these theorems make life hard for the realist in a certain precise sense, they in no way block realism with regard to quantum mechanics. Rather, the default assumption with regard to quantum mechanics, like any other scientific theory, is that the aim of a successful theory is to describe the physical world.

2.1 Quantum Mechanics as Incomplete

In the previous chapter I explained the difficulty of interpreting quantum theory. One response to this difficulty I mentioned is that quantum mechanics is *incomplete*—it is only a partial description of the physical world and needs to be supplemented in order to be fully satisfactory as a scientific theory.[1] The claim that quantum mechanics provides an incomplete representation of the physical world was first explicitly made by Einstein, Podolsky, and Rosen (1935); the argument has become known as the EPR argument.

The original EPR argument concerns entangled states for continuous quantities (position and momentum), but Bohm (1951, 614) showed that the argument can be made much more simply using entangled spin states like the one we considered in the last chapter:

$$|S\rangle = \frac{1}{\sqrt{2}}(|\uparrow_z\rangle_1|\downarrow_z\rangle_2 - |\downarrow_z\rangle_1|\uparrow_z\rangle_2). \tag{2.1}$$

We saw there that if both electrons in this state have their spin measured along the z-axis, the results never agree: Either you get spin-up for the first and spin-down for the second, or vice versa. Indeed, you can read this prediction off the way the state is written: The square of the coefficient on the $|\uparrow_z\rangle_1|\downarrow_z\rangle_2$ term gives the probability of getting spin-up for the first electron and spin-down for the second, and the square of the coefficient on the $|\downarrow_z\rangle_1|\uparrow_z\rangle_2$ term gives the probability of getting spin-down for the first electron and spin-up for the second. Hence, these probabilities are 1/2 each, as observed. The squares of the coefficients on the (absent) $|\uparrow_z\rangle_1|\uparrow_z\rangle_2$ and $|\downarrow_z\rangle_1|\downarrow_z\rangle_2$ terms give the probabilities of getting spin-up for both electrons and spin-down for both electrons, respectively; these coefficients are both zero, so the corresponding results are never observed.

An interesting algebraic fact about state $|S\rangle$ is that it takes the same form whatever direction we choose for the basic spin states. Earlier, we chose to write it in terms of the eigenstates of z-spin. But suppose that instead we choose to write it in terms of the eigenstates of w-spin, where w is an axis that makes an angle of 120° with z. Then it can be shown (via a little tedious algebra[2]) that even though most states look different when written in this new basis, $|S\rangle$ looks exactly the same:

$$|S\rangle = \frac{1}{\sqrt{2}}(|\uparrow_w\rangle_1|\downarrow_w\rangle_2 - |\downarrow_w\rangle_1|\uparrow_w\rangle_2). \tag{2.2}$$

Reading the results of measurements along the w-axis off the coefficients as before, this means that there is a chance of 1/2 of getting spin-up for the first electron and spin-down for the second, a chance of 1/2 of getting spin-down for the first and spin-up for the second, and zero chance of getting spin-up for both or spin-down for both. That is, whether measured along z or along w, the results of spin measurements in the same direction always disagree.

Now what about the spin *properties* of electrons? Quantum mechanics (as typically understood) says that the eigenstates represent the electrons as having the corresponding properties, so that $|\uparrow_z\rangle_1|\uparrow_z\rangle_2$ represents both electrons as having the spin-up property along z and so on. But state $|S\rangle$ is not an eigenstate of the spins of the individual electrons in any direction: It is a sum of distinct eigenstates whatever basis we write it in. As we saw in the previous chapter, standard quantum mechanics doesn't ascribe spin properties to noneigenstates. We might take quantum theory to be a complete representation of the state of a physical system—in which case when two electrons are represented by a state like $|S\rangle$, they simply lack spin properties. Or we might take quantum theory to be an incomplete representation, in which case electrons represented by states like $|S\rangle$ have spin properties, even though quantum theory doesn't tell us what those properties are. The EPR argument claims to establish the latter.

The argument rests on the following metaphysical principle, described as a "criterion of reality": "If, without in any way disturbing a system, we can predict with certainty (i.e., with probability equal to unity) the value of a physical quantity, then there exists an element of physical reality corresponding to this physical quantity" (Einstein, Podolsky, & Rosen, 1935, 777). By "an element of physical reality," they mean a physical property of the system concerned. In the present case, this principle provides us with a sufficient condition for an electron to have a spin property, namely if we can predict with probability 1 the spin we will obtain on measurement. Clearly the eigenstates satisfy this principle: When a pair of electrons is described by state $|\uparrow_z\rangle_1|\uparrow_z\rangle_2$, for example, then we can predict with certainty that each will be found on measurement to be spin-up along z, and the theory says that there are elements of physical reality—spin properties of the electrons—corresponding to the quantum state. But the principle says nothing about states like $|S\rangle$ in which we cannot predict with certainty what result we will get when we measure the spins of the electrons; EPR don't want to beg the question and simply assume that electrons in such states either have or lack physical spin properties.

Now consider what happens if we begin with an electron in state $|S\rangle$ and then measure the spins of the two electrons sequentially: We measure the spin of the first electron along z, and then we measure the spin of the second electron along z. We can't predict the result of the first measurement with certainty: Given the state, there is a probability of 1/2 that we will get spin-up and a probability of 1/2 that we will get spin-down. But once we have the result of the first measurement in hand, we can predict the result of the second measurement with certainty, since it has to be the opposite of the first. That is, if we got spin-up for the first electron, we can be certain that we will get spin-down for the second, and vice versa. Applying the EPR criterion of reality entails that after the first measurement, at least, the second electron has a spin property.

So can we conclude that $|S\rangle$ is a state in which the second electron has a physical spin property? Not yet: the measurement of the first electron involves a physical interaction with the electrons, and perhaps this interaction changes the state from $|S\rangle$ to a different one in which the second electron has a spin property. The EPR criterion of reality requires that we predict the spin of the second electron without disturbing it, and perhaps the measurement on the first electron constitutes a disturbance of the second electron. After all, the measurement postulate of standard quantum mechanics says that when the spin of the first electron is measured, the state of the system "collapses" to an eigenstate of z-spin for the first electron, and it looks like such a collapse will cause the second electron to be in an eigenstate of z-spin, too. That is, if the first electron is measured to be spin-up, then the second term in state $|S\rangle$ disappears, leaving the system in a state in which the second electron is in a z-spin eigenstate, too—a state in which it is spin-down. The second electron acquires a spin when the first one does.

But EPR argue that a change of state for the second electron can be ruled out. This is because the pair of electrons can be spatially separated before either of them is measured. As Einstein appreciated more than most, special relativity rules out any causal influence traveling faster than the speed of light (see Chapter 5). So if the electrons are separated, the measurement performed on the first electron cannot instantly affect the state of the second. But then what about the collapse postulate? Recall that the quantum theory we have been looking at is an explicitly nonrelativistic theory, and the collapse postulate makes its nonrelativistic nature apparent, since it allows an intervention on one particle to instantly affect the state of a distant particle. So we know that the collapse postulate cannot be exactly right for spread-out systems.[3]

The upshot of the EPR argument is that since an observer of the measurement on the first electron can instantly predict with certainty the result of a measurement on the second electron, the second electron must already have a physical spin property, even though its state remains unchanged. Hence, the second electron has a physical spin property, either spin-up along z or spin-down along z, when its state is described by $|S\rangle$. Because quantum theory does not ascribe spin properties to electrons in state $|S\rangle$, this already shows that quantum theory is incomplete. But so far the incompleteness does not seem so problematic. After all, we have quantum states available in which both electrons have z-spin properties, namely the eigenstates; perhaps state $|S\rangle$ just represents our ignorance about whether the true state of the electrons is $|\uparrow_z\rangle_1|\downarrow_z\rangle_2$ or $|\downarrow_z\rangle_1|\uparrow_z\rangle_2$. But EPR go on to rule out this possibility, too. All the measurements so far have been along direction z, and the spin properties we have been considering have been in this direction, but we can make exactly the same argument using direction w. That is, the result of a measurement of the first electron along w allows us to predict with certainty the spin of the second electron along w, even though the measurement cannot have affected the state of the second electron. Consequently, state $|S\rangle$ must be one in which the second electron has a physical spin property, either spin-up or spin-down, along w as well.

But now quantum mechanics is in trouble. It looks like EPR have established that the second electron is either spin-up or spin-down relative to z and either spin-up or spin-down relative to w.[4] It is not just that quantum theory does not include these properties in its representation of the physical state—it *cannot* do so, given its representational machinery. States $|\uparrow_z\rangle_1|\downarrow_z\rangle_2$ and $|\downarrow_z\rangle_1|\uparrow_z\rangle_2$ are states in which the electrons have spin properties along the z-axis, but they are not states in which the electrons have spin properties along the w-axis. Likewise, the states in which the electrons have spin properties along the w-axis are not states in which they have spin properties along the z-axis. Given the structure of the vector space, it is easy to see that there cannot be any quantum state that allows you to predict with certainty the spin of the electron along both the z- and w-axes: No vector can point in two directions at once. So if the electrons in state $|S\rangle$ have spin properties for z and for w, then quantum mechanics cannot represent them, given the structure of the theory. We cannot say that $|S\rangle$ represents our ignorance about which quantum state corresponds to the physical properties in front of us; rather, the true physical properties must be represented by something extraneous to quantum theory, something that must be added to it. This is the sense in which EPR argue that quantum theory is incomplete.

Almost immediately, Bohr published a reply to the EPR paper in which he argues that quantum mechanics *does* provide a complete description of physical reality. Bohr takes aim at the EPR criterion of reality, claiming that it contains "an essential ambiguity" (1935, 696). The ambiguity, he claims, lies in the clause "without in any way disturbing a system": Bohr argues that there are two senses of "disturbance" at issue here. The first is a mechanical disturbance of the system, such as might be produced by a physical interaction with it (1935, 700). In this sense of "disturbance," the measurement on the first electron doesn't disturb the second, precisely because they can be widely separated. But there is, he contends, a second sense of "disturbance" at issue here: The measurement on the first electron constitutes "an influence on the very conditions which define the possible types of predictions regarding the future behavior of the system" (1935, 700). It is certainly true that the measurement on the first electron affects what we can predict about the second electron: If we measure the spin of the first electron along z, then we can predict the spin of the second along z but not along w, and if we measure the first along w, then we can predict the spin of the second along w but not along z. But what does this have to do with the physical properties of the second electron? Bohr claims that "these conditions constitute an inherent element of the description of any phenomenon to which the term "physical reality" can be properly attached" (1935, 700). That is, we can't say that the second electron has a certain physical property unless conditions are such that we can *actually* predict its value. The measurement we choose to perform on the first electron—along z or along w—affects what we can actually predict concerning the second electron, and in this sense the measurement on the first constitutes a disturbance of the physical state of the second.

To put it slightly differently, the ambiguity in the EPR criterion of reality concerns whether we should interpret it as saying (1) that the second electron has a physical spin property along z if we can *actually* predict its spin along z, or (2) that the second electron has a physical spin property along z if we *could* predict its spin along z by performing the right measurement on the first electron. EPR don't distinguish the two readings and tacitly assume the latter. Bohr insists that the proper formulation for the quantum world is the former. If Bohr is right, then the last part of the EPR argument doesn't go through: We shouldn't say that $|S\rangle$ is a state in which the second electron has spin properties for the z and w directions simultaneously, because it is impossible to measure the spin of the first electron along both directions simultaneously. The most that can be said is that $|S\rangle$ is a state in which the second electron may have a spin property for one of the two directions. This allows Bohr to rescue a sense in which quantum mechanics might be said to be complete: Even though the theory doesn't specify the spin of the second electron along the z-axis when the state is $|S\rangle$, the theory does have the resources to represent the second electron as being spin-up along z or spin-down along z. Similarly for the w direction. As long as we don't have to represent the spin of the second electron along z and w simultaneously, the representational resources of quantum mechanics are sufficient.

But why think that Bohr is right—that the first reading of the EPR criterion of reality is correct? His position looks unmotivated: How can what we can *actually* predict affect the physical properties of the second electron, given that the measurement that allows us to make the prediction has no direct effect on the second electron at all? One possibility is to read his argument as espousing a kind of instrumentalism about quantum mechanics, so that "Quantum mechanics says that the electron is spin-up along z" just means "I can use quantum mechanics to predict that the electron is spin-up along z." Under this reading, what we can actually predict affects what properties things have because property talk is just an indirect way of talking about predictions. Indeed, Bohr is sometimes taken to be an instrumentalist, in part because of his personal connections to the logical positivist movement in Germany and Austria.

However, most contemporary commentators reject this reading. Bohr's relationship with the logical positivists was complicated (Faye, 2010), and a wholehearted instrumentalism cannot account for what Bohr actually says about quantum mechanics (Howard, 1994, 210; Halvorson & Clifton, 2002, 5). If quantum mechanics is just a predictive instrument, then an easy reply to the EPR argument is available: Quantum mechanics does not pretend to describe the physical world, so the project of trying to interpret it as descriptive is a fool's errand. But Bohr does not simply dismiss the EPR criterion of reality as empty metaphysics; he attempts to refine it and to interpret it such that quantum mechanics can function as a complete description of physical reality. This is not an instrumentalist project. Bohr's position is based on the particular structure of quantum mechanics, not on some global metaphilosophical position; his arguments are not applicable to other theories. To accuse Bohr of simply promulgating the positivism of his era is to overlook his actual arguments and to ignore the uniquely problematic nature of quantum mechanics.

So how are we to read Bohr's argument then? Bohr claims that the possession of physical properties needs to be reconceptualized for the quantum realm: Quantum mechanics requires "a radical revision of our attitude towards the problem of physical reality" (1935, 697). The radical revision he has in mind seems to be a kind of contextualism: What properties a system has depends not only on its quantum state but also on its physical environment.[5] In particular, Bohr claims that the z-spin of an electron is only defined when the physical environment of the system is such that we could actually measure its z-spin.[6] But the physical environment can never be such that we can actually measure z-spin and w-spin at the same time, since the measuring magnets need to be aligned along different directions in each case. So the z-spin and w-spin of an electron are never simultaneously defined. This allows Bohr to argue that quantum mechanics is complete as it stands: We do not need to supplement quantum mechanics so as to ascribe both a z-spin property and a w-spin property to a particle because a particle cannot have a spin property along two distinct directions simultaneously. If the physical context is such that the spin of the electron is defined along z, then it is undefined along w, and vice versa.

Bohr calls pairs of properties of this kind, where the definability of one precludes the definability of the other, *complementary* properties, and he makes complementarity—the existence of complementary properties—the heart of his interpretation of quantum mechanics (Howard, 2004). In one sense complementarity is undeniable: Some properties are simultaneously measurable and some are not.[7] But this does not mean that Bohr's interpretation of quantum mechanics is forced on us, or that Bohr is correct in his claim that quantum mechanics is complete as it stands. In order to make the claim that quantum mechanics is complete, Bohr has to reconceptualize physical properties as contextual—as dependent on the physical context of a system, not just on the system's intrinsic state. This is certainly a radical departure from our classical view of physical properties, as Bohr himself admits. It is not clear that such a conception of physical properties is metaphysically tenable: How, exactly, does the physical environment affect which properties a system has? Nor is it clear that appealing to such a conception solves the problem of ascribing properties to entangled states. Recall that a measurement on the first particle does not physically disturb the second electron; then how can a choice of spin direction for a measurement on the first electron affect which properties the second electron has? The choice of measurement direction for the first electron changes the physical environment of that electron, but it is hard to see how it can also change the physical environment of the second electron when they are spatially separated.

We will return to these questions later. But for the moment, the point is just that Bohr's response to the EPR argument is not decisive. EPR argue that quantum mechanics is incomplete, tacitly relying on an assumption that physical properties are not contextual. Bohr responds that quantum mechanics can be interpreted as complete if one is prepared to regard physical properties as contextual. But why go to such great lengths to be able to interpret quantum mechanics as complete? No theory is sacrosanct; why not supplement it, rather than countenance a radical revision in our metaphysics of properties? It turns out, though, that later theoretical developments substantially tip the weight of argument in Bohr's favor. These developments are the so-called no-go theorems.

2.2 No-Go Theorems

Let us suppose that the tacit assumption of the EPR argument is right and physical properties are not contextual—that is, the electrons in the entangled state $|S\rangle$ possess spin properties for any direction in which their spin *might* be measured, not only for those directions in which it is *actually* measured. Since quantum theory doesn't represent these spin properties, it is incomplete, and we need to add variables representing them in order to complete it.

As a particular example, consider spin measurements in three directions, v, w and z, lying in a plane 120° apart from each other. Then each electron must possess a spin property, either spin-up or spin-down, along v, along w, and along z. One straightforward way to represent these spin properties is as a triple of letters, where the first letter represents the spin along v, the second the spin along w, and the third the spin along z. So, for example, *uud* represents the electron as spin-up in the v direction, spin-up in the w direction, and spin-down in the z direction, as shown in Figure 2.1. The inclusion of these spin properties constitutes a step toward the completion of quantum mechanics that EPR called for. The EPR argument concludes that we can't explain our observations on an entangled pair of electrons without such properties.

So how should we explain our observations *with* these spin properties? Recall what the observations are. As shown earlier, if the electrons have their

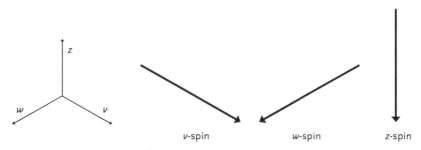

FIGURE 2.1 Spin properties of a *uud* electron.

spins measured along the same direction, the results never agree. But as shown in Chapter 1, if they have their spins measured along directions 120° apart, the results agree 3/4 of the time. Explaining these results looks like a simple problem of assigning the right combination of spin values to the electrons. First, we explain the fact that when the two electrons have their spins measured in the same direction the results never agree. This is easy: The spin properties of the two electrons must be the opposite of each other for each possible measurement direction. For example, if the spin properties of the first electron are *uud*, those of the second electron must be *ddu*. So given this initial constraint, the possible spin properties of the two electrons are as shown in the following table:

	Electron 1	Electron 2
1	*uuu*	*ddd*
2	*uud*	*ddu*
3	*udu*	*dud*
4	*udd*	*duu*
5	*duu*	*udd*
6	*dud*	*udu*
7	*ddu*	*uud*
8	*ddd*	*uuu*

Now let us turn to explaining the fact that when the electrons have their spins measured in directions 120° apart the results agree 3/4 of the time. Since directions *v*, *w*, and *z* are all 120° apart, what we need is for the spin properties appearing in different elements of the triple for each electron to agree 3/4 of the time. So for example, the first element for electron 1 needs to agree 3/4 of the time with the second element for electron 2, since the first element corresponds to direction *v* and the second element to direction *w*. How can we ensure that?

Clearly if the spin properties of the two electrons are as in the top row of the table, the results of measurements in different directions will never agree. The same goes for the bottom row. What if the possessed values are as in row 2? Well, there are six distinct ways of measuring the two electrons along different directions—*vw*, *vz*, *wz*, *zw*, *zv*, and *wv* (where the first letter is the measurement direction for the first electron and the second letter the direction for the second electron). For two of these possible measurements (*vw* and *wv*), the spin properties of the two electrons differ. For the remaining four (*vz*, *wz*, *zw*, and *zv*), the spin properties of the two electrons are the same. And exactly the same is true for the spin properties in rows 3–7: Two out of six possible measurements yield different results, and four out of six yield the same result.

So how can we assign spin properties to the electrons to make sure that their spins agree 3/4 of the time for different measurement directions? If the measurements performed on the electrons can depend on their spin properties, then the problem is relatively easy to solve.[8] But presumably this kind of scheme is impossible to instantiate, because any mechanism whereby the particle properties might influence the measurement directions could be easily undermined, for example, by picking the measurement directions at random. Similarly, if the spin properties of the electrons can depend on the measurements to be performed on them, then the problem is relatively easy to solve.[9] But presumably this kind of scheme is also impossible to instantiate, because there is no way that *future* measurements can influence an electron's *current* physical properties. So the spin properties can't influence the measurement directions, and the measurement directions can't influence the spin properties. In other words, the spin properties of the electrons and the measurements to be performed on them are independent of each other; call this the *independence assumption*. Finally, if the measurement outcome for the first electron can influence the spin properties of the second electron, the problem is relatively easy to solve.[10] But presumably this kind of scheme is impossible, too, because the two measurements can be as close together in time and as far apart in space as we like, so the influence would have to be infinitely fast. In other words, the result of the first measurement cannot affect the properties of the second electron; call this the *locality assumption*.

Given these two assumptions—independence and locality—if the properties of the electrons are chosen from those in rows 2–7, then for measurements in different directions the results will agree 2/3 of the time—since for these rows 4 out of 6 possible measurements yield results that agree, and the independence assumption means that the measurement directions are independent of the spin properties. If the spin properties are chosen from all eight rows, the results will agree less than 2/3 of the time, since rows 1 and 8 never produce agreement. So there is no strategy for assigning properties to electrons that can yield better than 2/3 agreement for measurements in different directions. Our simple problem of assigning spin properties to electrons to reproduce the observed results turns out to be impossible! This is Bell's theorem (Bell 1964): Given our two plausible assumptions, no assignment of spin properties to the two particles can reproduce the measurement results we observe—the results correctly predicted by quantum theory.[11]

Bell's theorem is perhaps the most celebrated of the no-go theorems, but a second theorem, proved by Kochen and Specker (1967), is arguably equally important.[12] Their theorem is more complicated to prove, but it relies on fewer assumptions than Bell's theorem. Both theorems engage the same question—whether the predictions of quantum mechanics can be reproduced by a theory that supplements quantum mechanics with extra physical properties. Bell considers a special quantum state—an entangled state of a pair of particles—and just a few measurable properties of those particles—their spins along three different directions. He asks whether the quantum mechanical predictions concerning these spins can be reproduced by assigning physical spin properties to the electrons, and he shows that given locality and independence, the answer is negative. Kochen and Specker consider a general quantum mechanical system and ask whether *every* quantity for which quantum mechanics makes predictions can simultaneously be ascribed a physical property consistent with that prediction. Like Bell, they show that the answer is negative; more precisely, they show that any attempt to reproduce the predictions of quantum mechanics using physical properties will result in ascribing contradictory physical properties to the system. To generate the contradiction, Kochen and Specker use a three-dimensional vector space—that is, a system with three distinct basic states—and a set of 117 potential measurements on the system. Given the number of potential measurements required to generate a contradiction, the proof is rather complicated, so I spare you the details here.[13] The proof generalizes to any system with three or more distinct basic states.

Bell's theorem concerns two measurements, one on the first electron and one on the second. Although the Kochen-Specker theorem invokes 117 potential measurements, only one of these measurements is made; the problem concerns how to arrange the underlying physical properties such that the predictions of quantum mechanics are reproduced whichever of the 117 possible measurements is chosen. Because only one measurement is involved, there is no need to worry about one measurement disturbing the properties detected by another measurement, so the Kochen-Specker theorem is not subject to a locality assumption. However, it is subject to an independence assumption: Kochen and Specker must assume that which measurement is chosen has no effect on the physical properties of the system, or else it is easy to reproduce the quantum mechanical predictions using physical properties.

2.3 What Do the Theorems Prove?

The no-go theorems are often regarded as entailing that the completion of quantum mechanics that EPR envisioned is simply not consistent with our empirical observations. For this reason, they are often referred to as "no hidden variable theorems"—theorems that rule out the completion of quantum mechanics by adding extra properties described by hidden variables, where "hidden" just means "not appearing in the standard theory."[14] It is a short step from here to an antirealist conclusion: The EPR argument shows that quantum theory must be completed if it is to describe the physical world, and the no-go theorems show that no such completion is possible, so quantum theory *cannot* describe the world, however we supplement it.

If this argument is sound, then quantum theory cannot be literally true of the domain it purports to describe because *no* theoretical description of the microphysical domain is possible. What's more, this antirealist conclusion rests on empirical premises: It is the experimentally confirmed predictions of quantum mechanics itself that can't be accommodated within a descriptive theory. So we apparently have a new form of argument for antirealism here, one motivated not by global observations about the whole scientific enterprise, but by local empirical findings in a particular domain of inquiry.[15]

But is it sound? While the empirical predictions on which the no-go theorems are based are unassailable, each theorem, as we have seen, rests on other assumptions as well. These assumptions are plausible, but that doesn't mean they can't be challenged. Indeed Bell himself preferred to conclude that one of the assumptions on which his theorem is based must be false, noting that "what is proved by impossibility proofs is lack of imagination" (1982, 997). That is, Bell shared Einstein's conviction that the measurement results we observe must be explained by preexisting properties of the system concerned; he called these properties "beables," since he found the term "observables" used by most physicists to be too instrumentalist-sounding (Bell 2004, 52). He agreed with Einstein that these "beables" are not adequately described by quantum mechanics and need to be added to the theory. But in light of his own theorem, he was led to the surprising conclusion that one of his assumptions must be false. Various possible ways of denying these assumptions will be explored in subsequent chapters.

In particular, the independence assumption is a premise in both Bell's theorem and the Kochen-Specker theorem. Independence is essentially an assumption that the physical properties in question are noncontextual. As noted in the previous section, choosing what measurement to perform on a system is tantamount to choosing the physical environment of the system. So if physical properties depend on the physical context of a system as well as its intrinsic state, then they can depend on the measurement performed on the system. Taken at face value, the lesson of the Kochen-Specker theorem seems to be that if there are any physical properties responsible for the measurement results predicted by quantum mechanics, they must be contextual properties. So on this issue, at least, it looks like Bohr was right: While we might like to think that there are noncontextual physical properties underlying and explaining the results of our measurements, such properties are simply not compatible with the empirical evidence.

But, of course, the debate between Einstein and Bohr was not primarily about contextuality: It was about the completeness of quantum mechanics as a descriptive theory. If physical properties are contextual, then the last step of the EPR argument does not go through: There is no need to ascribe preexisting properties to a system corresponding to every measurement that might be performed on it, since the measurement plays a role in determining the properties. But even so, it does not follow that the quantum mechanical description of a system is complete. Consider the entangled state $|S\rangle$ again and suppose that both electrons have their spins measured along the z-axis. This fixes the physical context of the system, and so according to Bohr fixes the physical properties of the system. So what are these properties? Quantum mechanics doesn't say: Nothing in the theory tells us whether the electrons in this situation are spin-up or spin-down, only that their spins are the opposite of each other. Instead, quantum mechanics gives us probabilities via the Born rule: There is a 50% chance that electron 1 is spin-up and electron 2 spin-down, and a 50% chance that electron 1 is spin-down and electron 2 is spin-up. But how should we interpret these probabilities? If they represent our ignorance of the true state, this entails that there is a true state that quantum mechanics fails to describe. If they represent some kind of objective chance of each result coming about, this entails that there is a physical process that brings one of them about, and again, quantum mechanics fails to describe this physical process.[16] Either way, quantum mechanics is incomplete as a description of the physical system. Furthermore, contextuality itself is a puzzle. How exactly does the physical context of a system determine its physical properties? Again, it seems that Bohr has merely drawn our attention to the descriptive inadequacy of standard quantum mechanics, rather than resolve it. So even if the no-go theorems prove Bohr right about contextuality, we have our work cut out more than ever in understanding how the world can be such that quantum mechanics is true of it.

2.4 Rescuing Realism

The no-go theorems undoubtedly make life hard for the realist. Given some plausible assumptions, it seems to follow that no description of the microworld could possibly account for the experimental outcomes we observe. So if we want to hang on to the realist view that scientific theories aim at literal, descriptive truth, we have to reject one of these assumptions. But which one should go? Working out the pros and cons of the various alternatives makes up the majority of the rest of the book. For now I simply run through the major options to establish that there are at least realist strategies to be pursued.

Initially, it looks like independence has to be the assumption that must go, since both theorems rely on it, and I didn't appeal to any other assumption in my (admittedly cursory) exposition of the Kochen-Specker theorem. And indeed there is a realist program that seeks to complete quantum mechanics by adding properties that violate independence (Price, 1994). However, this program is somewhat of a minority approach, because finding a way for the properties of a system to influence the measurements to be performed on it, or vice versa, requires some fairly radical revisions to our standard picture of the causal structure of the world. Either we have to admit that future events (choices of measurement) can causally influence present properties, or we have to postulate some unsuspected common cause of both the system's properties and the measurements to be performed on it. Neither of these options looks immediately attractive, although we will see in later chapters that this approach has some advantages that may offset these costs.

So are there other ways around the Kochen-Specker theorem? Yes, because the proof also relies on several tacit assumptions. One of these is the assumption that every quantum mechanical operator corresponds to a set of physical properties of the system: There are z-spin properties and w-spin properties and so on, so that every measurement simply reveals a preexisting possessed property. It is possible to deny this assumption. For example, perhaps there are no z-spin properties or w-spin properties, and the results we obtain when we perform spin measurements can be explained in terms of the *position* properties of the constituents of the system.[17] The key point is that there may not be a simple one-to-one correspondence between measurement outcomes and these underlying possessed properties. For example, it could be that a given configuration of underlying position properties could result in either a spin-up outcome or a spin-down outcome for a z-spin measurement, because the dynamical laws obeyed by the underlying constituents are indeterministic. In that case, the measurement results are explicable, even though there isn't a distinct preexisting property for each outcome. Without distinct preexisting properties for each possible measurement outcome, the Kochen-Specker construction doesn't succeed at ascribing *contradictory* properties to the system in question.

The major realist approaches to quantum mechanics all take advantage of this loophole in the Kochen-Specker theorem, but they do so in distinct ways. Spontaneous collapse approaches take the route just mentioned: They take position properties as basic and invoke an indeterministic law governing them. Bohmian hidden variable approaches also take positions to be basic, but in this case the law governing the position properties is deterministic. However, the law is such that the outcome of a spin measurement can depend on which way up the measuring device is placed, so again the preexisting properties of the particle itself don't determine the spin result.[18] Many-worlds approaches entail that a z-spin measurement performed on a state that is not a z-spin eigenstate results in *both* possible results, spin-up and spin-down, each existing in its own "branch" of reality. So again, there is no preexisting property of the system that determines the unique outcome of the measurement, in this case because the measurement has no unique outcome.[19]

These three approaches will be formally introduced in Chapter 3. What they have in common is that measurement results are explained without appealing to a separate preexisting property for each possible outcome. This takes care of the Kochen-Specker theorem. However, it doesn't automatically deal with Bell's theorem. Bell's theorem concerns spin properties, so one might think that one can get around it in the same way, namely by avoiding the ascription of genuine spin properties to particles. But though the spin results for each individual particle can be explained in terms of underlying properties other than spin, this strategy doesn't immediately help explain the correlations between the outcomes for the two particles. Suppose, for example, that spin results are explained in terms of underlying positions obeying an indeterministic law; an indeterministic law operating separately for each particle doesn't explain why measurements of z-spin for the two particles always disagree. Or suppose that the measurement result for each particle depends on which way up the measuring device for that particle is placed; again, since there need be no connection between which way up the two devices are placed, this doesn't explain why measurements of z-spin for the two particles always disagree.

Spontaneous collapse approaches and Bohmian hidden variable approaches deal with these correlations in the same way, namely by denying Bell's locality assumption. Bell assumed that a measurement on one particle doesn't affect the state of the other, but the laws invoked by both these approaches to govern the underlying position properties incorporate precisely this form of action at a distance. Hence, they can account for the correlations between the two measurement outcomes by supposing that a measurement performed on one particle directly influences the properties of the other.[20] Many-worlds approaches, on the other hand, *can* get away with the same kind of response to both the Kochen-Specker theorem and Bell's theorem. Recall that the many-worlds response to the Kochen-Specker theorem is that there is in general no unique measurement outcome, and so no need for a unique possessed property to explain it. The same goes for Bell's theorem: The measurement on each particle yields *both* a spin-up outcome and a spin-down outcome, in two distinct branches of reality. The correlations between the results are explained by the fact that for measurements in the same direction, there is one branch of reality containing a spin-up result for the first particle and spin-down for the second, and one branch of reality containing spin-down for the first particle and spin-up for the second. Hence, there is no need to postulate unique properties for each direction in which the particles' spins might be measured. That is, both the Kochen-Specker theorem and Bell's theorem rely on a tacit uniqueness assumption—that a measurement always has precisely one outcome—and many-worlds approaches to quantum mechanics violate this assumption.

So it looks like we have plenty of realist options available. The various strategies sketched here (to be explored in detail later) each provide a way to construct a description of the microphysical world that explains the empirical results of quantum mechanics. Bell's theorem and the Kochen-Specker theorem show that there are important and perhaps counterintuitive constraints that such descriptions must obey, but they do not rule them out altogether.

2.5 Conclusion

The considerations of the previous section show that while the no-go theorems make life difficult for the realist, they do not make life impossible.

The theorems rule out a descriptive account of quantum phenomena only on the basis of certain (explicit and tacit) assumptions, and these assumptions can be denied. But why bother? As we shall see in subsequent chapters, each of the realist strategies previewed here introduces its own range of problems and challenges. Why not rest content with quantum mechanics as a good instrument for predicting measurement outcomes and give up on the project of describing the microphysical world?

Put in such stark terms, the answer is obvious: It is the business of science to describe the world, and this is a difficult enterprise. Unforeseen challenges in providing an empirically adequate description do not warrant giving up. Of course, if describing the world is *impossible*, then that's a different matter. But this is not the situation we face in understanding quantum mechanics—at least, not yet.[21] If, at the end of the day, it turns out that none of these realist strategies is tenable, then absent some radical new theoretical perspective on quantum phenomena, we might have to conclude that we cannot describe the quantum realm. But until that point, the project of pursuing a literally true account of the quantum world remains a viable one. In the next chapter, we will look at several ways this project might play out.

3| Underdetermination

So what does quantum mechanics tell us about the nature of the physical world? In the previous chapter we considered the worry that it tells us nothing because quantum mechanics shouldn't be construed as a descriptive theory. We found this worry unsubstantiated: Although any descriptive account of the microworld must violate certain plausible assumptions, several such approaches are available. In this chapter we will consider the opposite worry: Rather than there being no description of the microworld available, it looks like there may be several competing descriptions, offering different stories about the constituents of the world and their behavior.

This is a familiar worry. Underdetermination has long been considered a barrier to drawing metaphysical conclusions from scientific theories. If a well-confirmed theory describes the world in a certain way, then it is always possible to construct a theory that is equally well confirmed but describes the world in a different and incompatible way. One might be led by such considerations to conclude that trying to learn anything of a metaphysical nature from our scientific theories is a waste of time (van Fraassen, 1980, 69).

Formally speaking, it is true that one can always construct an alternative to a given theory with a different structure but the same empirical consequences. It is not clear, though, to what extent such hypothetical underdetermination threatens scientific realism. These purpose-built alternatives are typically gerrymandered, ugly, unusable, and disconnected from other theories we accept. In short, no scientist would ever take them seriously. Realists typically argue that what scientists take seriously is not just a matter of taste or historical contingency; rather, scientists' judgments concerning "theoretical virtues" such as simplicity, utility, and consilience with the rest of science are important because these virtues carry evidential weight (Boyd, 1973; Psillos, 1999, 171). If that is right, then the existence of specially constructed alternatives is no threat to the claim of the original theory to correctly describe the world: The alternatives do not have the theoretical virtues of the original theory, and hence their descriptions of the world are not genuine competitors.

Whatever the merits of this realist response in general, the particular case of underdetermination presented by quantum mechanics is untouched by this argument. The three main versions of quantum mechanics to be explored in this chapter are not mere hypothetical constructs, but genuine scientific theories, proposed, discussed, and compared in the physics journals. Their simplicity, usability, and consilience with the rest of science may be to some extent debatable (and some of these debates will be taken up in the rest of this book), but unlike constructed alternatives, they are certainly not obviously devoid of the theoretical virtues. So the incompatible descriptions of the world they present us with do seem to be genuine competitors (Barrett, 2003, 1211).

This makes quantum mechanics a particularly fascinating case study in underdetermination, one that is possibly unique in the history of science. There have been competing theories, of course, but typically they make different predictions; we may not know right now which description of the world is correct, but we expect one of them to win out in relatively short order.[1] It looks like no such appeal is possible here, and even an appeal to the broader evidential context of the theoretical virtues is not immediately decisive. Does that mean that the skeptical argument wins the day—that at least in the case of quantum mechanics, no metaphysical conclusions can be drawn from our physical theories? I argue here that it does not, although we may have to qualify our metaphysical claims somewhat. First, though, let me lay out the fundamental problem with textbook quantum mechanics, suggested but not made precise in Chapter 1, that gives rise to these three competing theories.

3.1 The Measurement Problem

Recall from Chapter 1 that the theory of quantum mechanics has three elements: the quantum state, the dynamical law, and the measurement postulate. The last element may have struck you as odd: Every physical theory involves some kind of representation and some kind of dynamical law by which this representation changes over time, but there doesn't seem

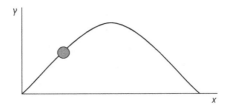

FIGURE 3.1 Trajectory of a baseball.

to be any analog of the measurement postulate in other physical theories. Consider, for example, a calculation using classical mechanics to determine how far a baseball thrown at $v = 10\,\text{m/s}$ at $45°$ above horizontal will travel before it hits the ground. We can represent the position of the baseball using the coordinates (x, y), where x is the horizontal distance traveled and y is the height of the baseball above the ground. Since the acceleration due to gravity is downward and of magnitude $g = 9.8\,\text{m/s}^2$, a little algebra shows that the baseball follows the parabolic curve shown in Figure 3.1, with coordinates given by $y = -(g/v^2)x^2 + x$. Solving for $y = 0$ tells us that the baseball hits the ground when it has traveled a horizontal distance of v^2/g, or $10.2\,\text{m}$. Note that I have not mentioned measurement at all: The final state of the baseball, represented by the coordinates $(10.2, 0)$, tells us where the baseball hits the ground, whether anyone measures it or not.

Compare this with a calculation using the matrix mechanics formulation of quantum mechanics. For example, suppose we want to know how an electron e that is initially in a superposition of distinct z-spin states will be deflected when it passes through a magnetic field oriented along the z axis. As we learned in Chapter 1, there are two eigenstates of z-spin for an electron, spin-up, written $|\uparrow\rangle_e$, and spin-down, written $|\downarrow\rangle_e$. (I have left out the subscript "z" on the arrows, since all the spins considered in this chapter are z-spins.) The situation for continuous quantities like position is a bit more complicated. As explained in Chapter 1, strictly speaking there are no position eigenstates among the possible quantum states of a system. However, if we divide up space more coarsely, into regions rather than precise points, then one can construct a discrete operator that does have well-defined eigenstates. For present purposes we can simplify matters by considering only the regions of space the electron enters when it is undeflected, deflected upward, and deflected downward. Then we can consider an operator with three eigenstates, $|0\rangle_e$, $|+\rangle_e$, and $|-\rangle_e$, corresponding to the electron being in these three regions of space. We can combine the spin states and position states into overall electron states, so that, for example, $|0, \uparrow\rangle_e$ represents an undeflected, spin-up electron.

To calculate what happens when an electron in a superposition of spin-up and spin-down states is passed through the magnetic field, we can appeal to the linearity of the quantum dynamics mentioned in Chapter 1. Recall that linearity says the following:

If $|\psi_1\rangle$ evolves to $|\psi_1'\rangle$

and $|\psi_2\rangle$ evolves to $|\psi_2'\rangle$ (3.1)

then $a|\psi_1\rangle + b|\psi_2\rangle$ evolves to $a|\psi_1'\rangle + b|\psi_2'\rangle$.

If an electron in the undeflected, spin-up state is passed through a magnetic field oriented along the z axis, it is deflected upward, and an electron in the undeflected, spin-down state is deflected downward. That is, an initial state $|0, \uparrow\rangle_e$ evolves to $|+, \uparrow\rangle_e$ and an initial state $|0, \downarrow\rangle_e$ evolves to $|-, \downarrow\rangle_e$. But suppose that the actual initial state of the electron is the superposition

$$\frac{1}{\sqrt{2}}(|0, \uparrow\rangle_e + |0, \downarrow\rangle_e). \tag{3.2}$$

Then the linearity of the quantum dynamics entails that it evolves to the final state

$$\frac{1}{\sqrt{2}}(|+, \uparrow\rangle_e + |-, \downarrow\rangle_e). \tag{3.3}$$

This final state is not a z-spin eigenstate, and neither is it a location eigenstate, so we cannot say based on this state what the electron's spin or location is. We cannot simply read the outcome off the final state, as we did in the baseball case. Rather, we have to apply the measurement postulate, which tells us that the probability of obtaining a particular outcome—of the state "collapsing" to a particular eigenstate—is given by the square of the coefficient on that eigenstate. In this case, if we were to measure the location of the electron, we would find it deflected upward with probability 1/2 and downward with probability 1/2. If we find it deflected upward, then the postmeasurement state of the electron is $|+, \uparrow\rangle_e$, and if we find it deflected downward, then its postmeasurement state is $|-, \downarrow\rangle_e$.

But how exactly does the measurement do this? Measuring the location of the electron does not require anything fancy; the simplest way is just to run it into a screen that lights up where it hits, such as an old-fashioned fluorescent TV screen. Indeed, we can model this process using quantum mechanics itself. The screen is made up of particles, and presumably there is a wide range of states of those particles such that the screen is blank, a disjoint range such that there is a flash in the top half of the screen, and yet another disjoint range such that there is a flash in the bottom half of the screen. For simplicity, let us consider only these three such ranges, setting aside those corresponding to other patterns of illumination on the screen. Then we can consider a discrete operator that has three eigenstates, $|0\rangle_s$, $|+\rangle_s$, and $|-\rangle_s$, corresponding to no flash, a flash in the top half, and a flash in the bottom half, respectively.

In terms of this notation, the initial state of the physical system consisting of the electron and the screen can be written

$$|0\rangle_s \frac{1}{\sqrt{2}}(|0,\uparrow\rangle_e + |0,\downarrow\rangle_e), \qquad (3.4)$$

or equivalently

$$\frac{1}{\sqrt{2}}(|0\rangle_s|0,\uparrow\rangle_e + |0\rangle_s|0,\downarrow\rangle_e), \qquad (3.5)$$

since we can distribute the initial term $|0\rangle_s$ inside the parentheses. If the screen is a reliable detector of electrons (which we will assume it is), then a spin-up electron will always produce a flash in the top half of the screen and a spin-down electron will always produce a flash in the bottom half of the screen. That is, the initial state $|0\rangle_s|0,\uparrow\rangle_e$ always evolves to the final state $|+\rangle_s|+,\uparrow\rangle_e$ and the initial state $|0\rangle_s|0,\downarrow\rangle_e$ always evolves to the final state $|-\rangle_s|-,\downarrow\rangle_e$. Then by the linearity of the quantum dynamics again, the actual initial state (3.5) evolves to the final state

$$\frac{1}{\sqrt{2}}(|+\rangle_s|+,\uparrow\rangle_e + |-\rangle_s|-,\downarrow\rangle_e). \qquad (3.6)$$

Note that (3.6) is not an eigenstate of the screen display operator, so we cannot say based on this state where the flash is. So incorporating the measurement process into our quantum mechanical analysis does not get us any closer to figuring out where the result of our measurement comes from. We can of course apply the measurement postulate again at this point: It says that if we measure the location of the flash, we will find it in the top half with probability 1/2 and in the bottom half with probability 1/2. But our quantum mechanical analysis of measurement does not allow us to *explain* the measurement postulate, since inclusion of the measuring device in the quantum mechanical analysis retains both terms in the state, and to get one result or the other, one term or the other would have to be eliminated.

So there is something deeply mysterious about the role of measurement in textbook quantum mechanics. In the baseball case, we can mention measurement if we like, but measurement plays no role in the physics itself.

In the quantum case, however, measurement plays a crucial and ostensibly ineliminable role: Measurement literally brings about the outcomes that we observe. The dynamical law of quantum mechanics (the Schrödinger equation) is linear, which means that any physical interaction with the electron, if it proceeds according to this law, must retain the two terms in the state in exactly the same proportions as in the initial state (3.5). Linear dynamical evolution can never eliminate one of these two terms, but we need to eliminate one of them if the experiment is to have an outcome. So there is no way that the Schrödinger equation can produce the observed outcome in the way in which the dynamical law of classical mechanics produces the observed position of the baseball. Rather, the measurement postulate introduces an *exception* to the dynamical law: Physical systems evolve according to the Schrödinger equation *except* during a measurement.

There are at least two problematic aspects to this. The first is that "measurement" is not a precise term; one would be hard-pressed to specify necessary and sufficient conditions for a process to count as a measurement. And yet whether or not a given process counts as a measurement has a direct impact on the physical evolution of the system concerned. This problem was expressed in a particularly striking way by Schrödinger (1935) in his famous cat thought experiment. Schrödinger envisions a cat shut up in a box with a device that releases poison if (and only if) a particular radioactive atom decays, essentially an elaborate and cruel way of measuring whether the atom has decayed. After an hour, according to the Schrödinger equation, the atom will be described by a superposition of two terms, one of which is an eigenstate of it having decayed, and the other of which is an eigenstate of it having not decayed. Because the former state evolves to a state in which the cat is dead, and the latter state evolves to a state in which the cat is alive, the linear dynamical law acting on the superposition of decayed and undecayed eigenstates necessarily produces a superposition of alive-cat and dead-cat eigenstates.

Presumably (although more on this in Chapter 4) we know by observation that cats are always either alive or dead. So at some point in the physical process between the decay of the atom and the observation of the cat, the measurement postulate must kick in and "collapse" the superposition to one term or the other, thereby producing the outcome we observe. But at what stage in the physical process, precisely, does the measurement occur? Is it when the device detects the radiation from the atom's decay? Or when the poison is released? Or when the cat dies? Or not until the experimenter opens the box to examine its contents? Quantum mechanics doesn't say. But it makes a physical difference when the measurement postulate applies, since the measurement postulate overrides the usual linear dynamics. Without a precise specification, then, of which physical processes count as measurements and which do not, the dynamical laws of quantum mechanics are incompletely specified.

This is the first way in which the measurement postulate is problematic. But there is also a second problematic aspect. What is a measurement? At bottom, it is a physical process like any other: Measuring devices are just hunks of matter, obeying the same physical laws as any other hunk of matter. But then it is simply impossible for the physical laws to differ between measurements and nonmeasurements, as the measurement postulate requires. As we just saw, if measuring devices are just physical systems obeying the linear Schrödinger dynamics, then the application of a measuring device to a system cannot cause an exception to the linear Schrödinger dynamics. Measuring devices must obey the Schrödinger dynamics, since every physical system does, but they must also violate the Schrödinger dynamics if they are to enact the measurement postulate. Quantum mechanics, understood as including the measurement postulate, is not just incomplete; it is inconsistent.

This twofold problem is the much-discussed *measurement problem* for quantum mechanics, first exposed by Schrödinger's cat thought-experiment, and brought into clear focus in the works of John Bell. Bell's conclusion is that reference to measurement "should now be banned altogether in quantum mechanics" (1990, 20): The measurement postulate has to go. But if we get rid of the measurement postulate, what should we replace it with? The three major interpretations of quantum mechanics are different attempts to answer this question. I begin with the spontaneous collapse approach. This is the most recent of the three approaches to be expounded here, but it is also the one that is closest in spirit to the original textbook version of quantum mechanics.

3.2 Spontaneous Collapse Theories

The role of the measurement postulate is to precipitate collapse—to provide a mechanism whereby the premeasurement superpositions that are required to explain phenomena like interference can evolve to postmeasurement eigenstates in which the measurement has a well-defined outcome. It would be nice if there were a mechanism—a modification to the Schrödinger dynamics—that could accomplish the same thing without the appeal to measurement. This idea was proposed by Pearle (1976) and worked out in detail by Ghirardi, Rimini, and Weber (1986). The original theory has become known as the GRW theory, and the GRW theory and its descendants are collectively known as *spontaneous collapse* theories.

The essence of the GRW theory is quite simple. Rather than ascribing collapse, mysteriously, to measurements as such, Ghirardi, Rimini, and Weber propose a precise new physical law that operates in addition to the Schrödinger dynamics to generate the collapse process. The new law is totally unlike the Schrödinger dynamics: Where the latter is linear, continuous, and deterministic, the new law is nonlinear, discontinuous, and fundamentally probabilistic. In fact, the new law is probabilistic in two ways. The first way is that according to this law, every particle has a small probability per unit time of undergoing a "hit" in which its state jumps discontinuously to a state that is much more localized. Whether or not a particle suffers a hit in the next second is not determined by the prior state of the system, but simply occurs at random with an average rate of 1 hit every 10^{16} seconds.

A particle that suffers a hit has its wave function multiplied by a narrow three-dimensional Gaussian function (bell curve) centered on a random point in space—a function that is large very close to this random point, but small further away. This is the second way in which the new law is probabilistic: The point on which the hit is centered is not determined by the prior state of the system, but rather the prior state determines the probability distribution for the point. Because we want to recover the Born rule for measurement outcomes, this probability distribution is given by the square of the prehit wave function amplitude.[2] The width of the Gaussian hit-function is 10^{-5} cm, and the posthit wave function is renormalized (multiplied by a constant) so that the integral of the squared wave function amplitude over the whole space remains 1.

Let's see how this law works. Suppose that the wave function of a single particle is initially distributed symmetrically between two disjoint regions A and B, as shown on the left in Figure 3.2. Now suppose the particle undergoes a GRW hit. Because the two regions are associated with equal squared-amplitude, the hit has a 50% chance of being centered on region A and a 50% chance of being centered on region B. If the hit is centered on region A, the resulting wave function distribution is as shown on the right in Figure 3.2. Now almost all the wave function amplitude is concentrated in region A—not quite all of it, because the Gaussian hit-function has "tails" going to infinity, so there is still a tiny wave function remnant in region B (not shown to scale!).

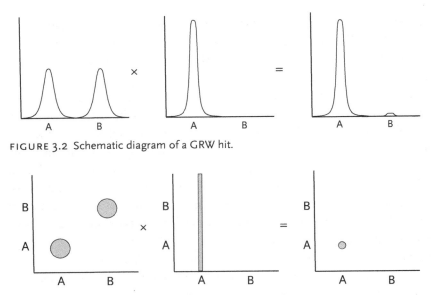

FIGURE 3.2 Schematic diagram of a GRW hit.

FIGURE 3.3 GRW hit for two correlated particles.

It is important to note that such collapses are incredibly rare: Since the hit rate is 1 per 10^{16} seconds, you can expect to watch a single particle for a hundred million years without seeing a hit. This is just as well; frequent collapses for individual particles would destroy interference effects, and interference effects are well-documented empirical phenomena. In fact, the Schrödinger dynamics is exceptionless in our observations of microscopic systems, so it is important that hits are rare enough to account for this.

But for macroscopic collections of particles, the hit rate becomes appreciable. A macroscopic object contains something of the order of 10^{23} particles, so the hit rate for the object as a whole becomes about 10^7 per second. Still, the hits during 1 second only affect 1 in 10^{16} particles; how does localizing a tiny proportion of the particles in an object localize the object as a whole? The answer is correlation. The particles in a solid object are strongly bonded together, so that when you move some of them, you move them all. In other words, the positions of the particles that make up a solid object are strongly correlated with each other. This means that a hit on any one of the particles that make up the solid object localizes all of them.

To see how this works, consider the two-particle system shown in Figure 3.3. The particles are each in a superposition of being in region A and being in region B, but they are also strongly bound together, so their positions are correlated. This is reflected in the wave function shown schematically in the figure. Recall from Chapter 1 that the wave function of a two-particle system resides in a six-dimensional space, with three dimensions for each particle. The diagram just depicts one dimension for each particle—the x direction for particle 1 along the horizontal axis and the x direction for particle 2 along the vertical axis—and the shaded areas represent regions of high wave function amplitude. So the wave function has two regions of high amplitude, one corresponding to both the particles being in region A, and the other corresponding to both the particles being in region B.

Suppose that one of the particles undergoes a hit. The hit function is a Gaussian in the coordinates of one of the particles, and constant in the coordinates of the other, so in our diagram it is represented by either a horizontal or vertical stripe, depending on which particle is hit. The diagram shows a hit on particle 1, centered on region A. Notice that due to the correlation between the particles, the hit also localizes almost all of the wave function amplitude within region A for particle 2.[3] So a hit on particle 1 results in both particles ending up in region A. Of course, for a pair of particles the hit rate is still too low to be observable, and similarly for superpositions involving single atoms or single molecules. But for a macroscopic solid object, this process means that if it ever ends up in a superposition of occupying two distinct locations, the GRW collapse process rapidly takes its state to one in which almost all its wave function amplitude is associated with one of the locations.

This is the GRW solution to the measurement problem. Suppose we have a particle that is initially in the state depicted in Figure 3.2 and we measure its location. To measure its location, we need to correlate its location with the position of a macroscopic object that we can see—a pointer on a dial or its functional equivalent. As we saw in the previous section, the linear Schrödinger dynamics results in a state in which the pointer is in a superposition of pointing at "A" on the dial and pointing at "B" on the dial—a state in which we cannot ascribe a position to the pointer. But the GRW collapse dynamics means that this superposition is unstable; in a tiny fraction of a second it collapses to a state in which it is pointing at "A" or to one in which it is pointing at "B," with probabilities given by the Born rule.

The key point here is that the references to measurement or a measuring device or a pointer are entirely eliminable; we could just say "the position of the particle is correlated with the position of a macroscopic solid object," and the result would be the same. Likewise, there is no need to worry that the microscopic-macroscopic distinction is vague: The new collapse law is entirely precise, and its consequences for any particular object can simply be derived. Microscopic systems can occupy superpositions of distinct locations (almost) indefinitely, and macroscopic systems are (almost) always very close to eigenstates of being in a particular location, and the combination of the Schrödinger dynamics and the GRW collapse dynamics can explain precisely why this is the case.

But a few problems remain. First, note that the postcollapse state for the particle in Figure 3.2 is not one in which *all* its wave function amplitude is located in region A, just one in which *most* of its wave function amplitude is so located. If the eigenstates for the location of the particle are the states in which all of the wave function amplitude is entirely contained in the relevant region, then the postcollapse state is not a location eigenstate, and we cannot say that the particle has a well-defined location even after the collapse. It might seem a simple matter to fix this—by making the link between property ascription and eigenstates more "fuzzy"—but this introduces some complications, considered at length in Chapter 4.

The GRW theory also walks something of an empirical knife edge. The theory introduces two new fundamental constants—the frequency of hits and the width of the hit function—and the values of these constants have to be carefully chosen to avoid conflicts with experiment. If the hit frequency is too high, microscopic superpositions would not have the stability they are observed to have, and if the frequency is too low, macroscopic objects would not acquire well-defined locations as quickly as they appear to.[4] If the hit function width is too large, then macroscopic objects would not be localized to the precision we can observe with our eyes, and if the width is too small, then objects would heat up to a measurable extent, because localizing objects increases their average kinetic energy. At the moment the experimental evidence leaves space for the GRW constants to occupy, but it is possible that future experimental developments, for example, in the production of long-lived superpositions involving many particles, will squeeze that space to zero.

As Albert notes, this business of choosing the values of the constants "may strike some readers as unpleasantly ad hoc" (1992, 99). But he continues "those readers certainly have their nerve," and his dismissal seems apt. It is not as if the standard theory of quantum mechanics has been falsified, and the GRW mechanism is an attempt to save quantum mechanics from falsification. Rather, the measurement problem shows that the standard theory is (at best) incomplete, and the GRW theory provides the relevant completion. The GRW proposal does what any reputable scientific theory does: It postulates a precisely defined mechanism to account for known data and leaves it up to the experimentalists to determine whether that mechanism is correct.

There are further, more serious problems with the GRW theory that will be taken up in later chapters, having to do with dimensionality (Chapter 7) and locality (Chapter 5). Let me complete this preliminary account of the GRW theory, though, by briefly considering its ontology. I formulated the theory in terms of the behavior of a system of particles, but if you consider the completed theory, you notice that the particles are nowhere in sight: An N-particle system is represented by a wave function in a $3N$-dimensional space. A more careful (but less intuitively accessible) presentation of the theory would be in terms of the wave function alone: The wave function evolves smoothly and deterministically for the most part, but occasionally it jumps discontinuously into a more localized state in three of its coordinates. For example, if you consider the phenomenon of two-slit interference described in Chapter 1, the GRW explanation for the phenomenon is that the wave associated with a single electron passes through both slits, the two components of the wave interfere, but when the electron is observed the GRW collapse process kicks in and concentrates the wave on a particular location. The GRW resolution to the "wave-particle duality" exhibited by quantum mechanics is that quantum reality is wave-like, but the wave sometimes bunches up like a particle. It is tempting to say, at least on a superficial reading, that according to the GRW theory there really are no particles.

Several variant spontaneous collapse theories have been developed since the GRW theory. A version has been developed in which the collapse process is continuous rather than consisting of discrete events (Ghirardi, Pearle, & Rimini, 1990). Some have suggested supplementing the wave function with a mass density distribution over space (Ghirardi, Grassi, & Benatti, 1995), and others have suggested that only the central point of each collapse event (rather than the entire wave function) should be considered real (Tumulka, 2006a). The latter two variants in particular will be important in later chapters.

3.3 Hidden Variable Theories

While the GRW theory treats particle phenomena as reducible to the behavior of waves, Bohm's theory (Bohm, 1952) takes particle phenomena to be irreducible. That is, Bohm's theory starts from the assumption that the flash on the screen in the two-slit interference experiment is caused by a genuine *particle* that travels from source to screen via one slit or the other, rather than by a wave that sometimes behaves as a particle. In particular,

Bohm's theory is formulated in terms of *point*-particles—entities that occupy a spatial point at any given time, and whose motions over time trace out lines in space-time (trajectories). Standard textbook quantum mechanics contains nothing corresponding to such objects. As noted in Chapter 1, there is nothing in either the vector or the wave function representation of states that can represent anything as having a precise position. So if we want to represent the positions of point-particles in our theory, that representation has to be *added* to standard quantum mechanics.

This is precisely what Bohm proposes. According to Bohm's theory, the complete representation of a quantum system consists of a wave function plus a set of coordinates representing the positions of the particles. So a one-particle system is represented by a wave function distributed over three-dimensional space, plus coordinates $\mathbf{x} = (x, y, z)$ that represent the position of a particle in that space. Likewise, a two-particle system is represented by a wave function distributed over a six-dimensional space, plus coordinates $\mathbf{x}_1 = (x_1, y_1, z_1)$ and $\mathbf{x}_2 = (x_2, y_2, z_2)$ representing the positions of two particles in three-dimensional space. Equivalently, the positions of the two particles can be represented as a single point $\mathbf{x} = (x_1, y_1, z_1, x_2, y_2, z_2)$ in the six-dimensional space occupied by the wave function, since a single point in the six-dimensional space represents the positions of two particles in three-dimensional space. The variable $\mathbf{x}$ is traditionally referred to as a "hidden variable"; hence, Bohm's theory and its descendants are collectively known as *hidden variable* theories.

According to Bohm's theory, the wave function evolves over time according to the Schrödinger dynamics; there are no collapses. But how do the particle positions change over time? What we want is to be able to interpret the squared wave function amplitude integrated over a particular region of space as the probability of finding the particles in that region; we want to recover the Born rule. So let us suppose that the particles are initially distributed over space in such a way that their probability distribution is the squared wave function amplitude; what we want is for them to move in such a way that their probability distribution *remains* the squared wave function amplitude, even as the wave function itself changes over time. Roughly speaking, this requires the particles to be attracted to regions of high squared amplitude and repelled from regions of low squared amplitude.

Slightly more precisely, what we want is to treat the probability distribution as flowing from place to place according to the Schrödinger equation, and then arrange for the particle positions to follow this flow. Suppose we are dealing with an N-particle system, and hence a $3N$-dimensional wave function $\psi(\mathbf{r}, t)$, where $\mathbf{r}$ represents the $3N$ spatial coordinates. Let $\rho(\mathbf{r}, t)$ be the probability distribution for the particles, and let $\mathbf{j}_n(\mathbf{r},t)$ be the flow of this distribution corresponding to particle n. Then what we want is to be able to define $\rho(\mathbf{r},t)$ and $\mathbf{j}_n(\mathbf{r},t)$ so that they satisfy the continuity equation

$$\frac{\partial}{\partial t}\rho(\mathbf{r},t) = -\sum_n \frac{\partial}{\partial \mathbf{r}_n} \cdot \mathbf{j}_n(\mathbf{r},t). \tag{3.7}$$

That is, we want the rate of change of the probability distribution in a given region to be the net flow of probability into that region. Bohm (1952, 169) shows how to define ρ and $\mathbf{j}_n$ to do just that. Because $\psi(\mathbf{r},t)$ is a complex-valued function, we can write it as $R(\mathbf{r},t)e^{iS(\mathbf{r},t)}$, where $R(\mathbf{r},t)$ and $S(\mathbf{r},t)$ are real-valued functions. Obviously, we want $\rho(\mathbf{r},t) = |\psi(\mathbf{r},t)|^2 = (R(\mathbf{r},t))^2$, so as to satisfy the Born rule. Less obviously, Bohm shows that if

$$\mathbf{j}_n(\mathbf{r},t) = \frac{(R(\mathbf{r},t))^2}{m_n} \frac{\partial}{\partial \mathbf{r}_n} S(\mathbf{r},t), \tag{3.8}$$

where m_n is the mass of the nth particle, then the continuity equation (3.7) follows from the Schrödinger equation. Then if the particle configuration $\mathbf{x}$ is initially distributed according to the probability distribution $\rho(\mathbf{r},t)$, it will remain distributed this way, provided that the particles move according to

$$\frac{d}{dt}\mathbf{x}_n = \frac{\mathbf{j}_n(\mathbf{x},t)}{\rho(\mathbf{x},t)} = \frac{1}{m_n}\frac{\partial}{\partial \mathbf{x}_n} S(\mathbf{x},t), \tag{3.9}$$

where $\mathbf{x}_n$ is the position of the nth particle.

The precise form of the law need not detain us, but a few features of it are worth remarking on. First, it is entirely deterministic: The initial wave function and the initial particle positions determine the wave function and the particle positions at all times. Second, it recovers the Born rule by design: When the positions of the particles are measured at the end of the experiment, their probability distribution is given by the squared wave function amplitude, because the initial distribution and the Bohmian law entail that this is always the case. Third, the measurement problem does not arise for Bohm's theory: The wave function always obeys the Schrödinger equation and never undergoes a collapse, and it is the particle positions rather than the wave function that generate the results of the measurement.

Figure 3.4 shows some of the possible Bohmian particle trajectories for a two-slit interference experiment—"some of them" because the possible

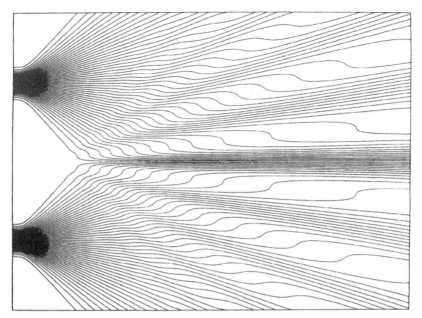

FIGURE 3.4 Bohmian particle trajectories for two-slit interference (from Philippidis, Bohm, & Kaye, 1982).

trajectories form a continuum. Compare the trajectories to the sketch of the wave function in Figure 1.2. Note how the trajectories bunch up in some places—where the wave function amplitude is high—and spread out in others—where the wave function amplitude is low. The particle follows one of these trajectories, and hence passes through one slit or the other. The wave function, on the other hand, passes through both slits, and the two components interfere. The resulting peaks and troughs in the wave function push the particle around according to the Bohmian dynamical law. So even though the particle passes through one slit, it is influenced by the wave function component passing through the other slit, and this explains why the probability distribution for the particle arriving at the screen when both slits are open is not just the sum of the distributions when one slit or the other is open.

Like the GRW theory, Bohm's theory faces serious problems having to do with dimensionality and locality, to be taken up in later chapters. Unlike the GRW theory, Bohm's theory (at least superficially) endorses a dual ontology: There are particles that determine the results of our measurements, and there are waves that push those particles around. Presumably the positions of the particles are referred to as "hidden variables" because before a measurement we have only statistical knowledge of their values. But as Bell (1981) has stressed, this name is misleading, since it is the state of the particles, not the wave function, that is directly manifest in the outcome of the measurement.

Since Bohm's theory, several variant hidden variable theories have been constructed. Bell (1987a) and Vink (1993) show how to take properties other than position to be the values of the hidden variables. Others, following the lead of van Fraassen (1979), take the properties that are ascribed precise values to vary according to the state of the system; these are the so-called *modal* hidden variable theories.

A final variant worth mentioning takes a different approach. Recall from Chapter 2 that Bell's no-go theorem rests on two assumptions, locality and independence. Bohm's theory and the variants mentioned so far evade the conclusion of Bell's theorem by violating locality: In equation (3.9) the motion of the nth particle depends on x—that is, it depends on the positions of *all* the particles in the system at the time in question, no matter how distant. But suppose instead we choose to violate the independence assumption—that is, we make the hidden variable values for a particle depend on the measurements that are yet to be performed on that particle. The most direct way to do that is by allowing the later measurement events to causally influence the earlier state of the particle (Price, 1994). This *retrocausal* hidden variable approach is described in more detail in Chapter 5.

3.4 The Many-Worlds Theory

Bohm's theory and the GRW theory each solve the measurement problem by replacing the measurement postulate with something else: The GRW theory replaces it with a physically precise collapse mechanism, and Bohm's theory replaces it with particles that have precise positions even when the quantum state is spread out. Everett (1957) takes a radically different approach to the measurement problem; he simply jettisons the measurement postulate and argues that the result is a complete and adequate theory.

This looks immediately problematic. Consider, for example, a single electron in a superposition of z-spin states:

$$\frac{1}{\sqrt{2}}(|\uparrow\rangle_e + |\downarrow\rangle_e). \tag{3.10}$$

As we saw earlier, if we pass such an electron through a magnetic field and then run it into a fluorescent screen, the state of the system consisting of the electron and the screen becomes

$$\frac{1}{\sqrt{2}}(|+\rangle_s|+,\uparrow\rangle_e + |-\rangle_s|-,\downarrow\rangle_e). \tag{3.11}$$

This state is a superposition of two eigenstates: $|+\rangle_s|+,\uparrow\rangle_e$, corresponding to the electron being spin-up, deflected upward, and hitting the upper part of the screen, and $|-\rangle_s|-,\downarrow\rangle_e$, corresponding to the electron being spin-down, deflected downward, and hitting the lower part of the screen. Consequently, state (3.11) cannot be straightforwardly interpreted as one in which the electron hits the upper part of the screen. Neither can it be straightforwardly interpreted as one in which the electron hits the lower part of the screen. Without the measurement postulate, the question of how we should interpret states like this becomes even more pressing.

There are two obvious interpretive options here (although neither is very promising at first glance). The first is to claim that (3.11) is *neither* a state in which the electron hits the upper part of the screen *nor* a state in which the electron hits the lower part of the screen. This option, which takes the outcome of the experiment to be *indeterminate*, is explored in Chapter 4. The second option is to claim that (3.11) is *both* a state in which the electron hits the upper part of the screen *and* a state in which the electron hits the lower part of the screen. This is the option that Everett pursues.

Everett calls his theory the "relative state" theory. What he wishes to bring to our attention is the internal structure of state (3.11): Relative to the first term, in which the electron is spin-up, the flash occurs in the upper half of the screen, and relative to the second term, in which the electron is spin-down, the flash occurs in the lower half of the screen. What Everett is suggesting is that we take the actual properties of a system to be represented by the relative states, not by the quantum state as a whole. That is, the electron has the property of being spin-up *and* the property of being spin-down, and the screen has the property of displaying a flash in the upper half *and* the property of displaying a flash in the lower half. The quantum state (3.11) tells us how these properties are related to each other: The flash is in the upper half relative to the electron being spin-up, and the flash is in the lower half relative to the electron being spin-down, but both related pairs are actually instantiated.

The obvious problem, of course, is that some of these properties are mutually exclusive: An electron can't be both z-spin-up and z-spin-down, and it can't be both deflected upward and deflected downward. What's more, we know empirically that when such experiments are conducted, there is always a single flash either in the upper half of the screen or the lower half, never both. How can Everett's theory account for this?

Consider what happens when an observer looks at the screen. Let $|+\rangle_o$ be a state of the observer in which she sees a flash in the upper half of the screen, and let $|-\rangle_o$ be a state in which she sees a flash in the lower half of the screen. Then by exactly analogous reasoning to that rehearsed for state (3.11), the final state of the system consisting of the electron, the screen and the observer is

$$\frac{1}{\sqrt{2}}(|+\rangle_o|+\rangle_s|+,\uparrow\rangle_e + |-\rangle_o|-\rangle_s|-,\downarrow\rangle_e). \tag{3.12}$$

According to Everett's theory, the observer sees both a flash in the upper half of the screen and a flash in the lower half of the screen. You might think that this even more directly demonstrates the empirical inadequacy of Everett's theory: Surely we know that observers have unique experiences! However, the relations between the observer's observations and the properties of the observed system are important: The observer sees a flash in the upper half relative to there being a flash in the upper half and the electron being spin-up, and sees a flash in the lower half relative to there being a flash in the lower half and the electron being spin-down. The structure of state (3.12) essentially partitions the properties into two sets. The properties in each set interact with each other in the standard physical ways—particle properties produce measuring device outputs and thence experiences—but there is no such interaction between the properties in different sets.[5]

This is suggestive. Everett's own exposition is brief,[6] but others have taken his central insight and developed it in their own way. In particular, several subsequent commentators note that the noninteraction of the two sets means that each set of properties behaves like its own *world*: There is the world in which the electron is spin-up, the flash is in the upper half of the screen and the observer sees the flash in the upper half, and there is the world in which the electron is spin-down, the flash is in the lower half of the screen and the observer sees the flash in the lower half.[7] These worlds are not mere possibilities, one of which is actual: the current proposal explicitly eschews adding anything like hidden variables to the theory that could distinguish the actual world from the mere possibilities. So the worlds are on an ontological par: if one is actual, they all are. This way of understanding Everett's approach is known as the *many-worlds theory*.

A case can be made along these lines that (a version of) Everett's theory is empirically adequate after all.[8] Every observer—including you and I—lives in a single world, and in each world the observed properties are perfectly well defined. When I see a flash in the upper half of the screen because the electron is spin-up, I know (if the many worlds theory is true) that there is another, equally real world in which my counterpart sees a flash in the lower half of the screen because the electron is spin-down. The lack of interaction between the worlds explains my lack of direct awareness of the measurement outcome in the other world; I only know about it because my theory tells me it is there.

Of course, it is not always the case that the two terms in a superposition fail to interact. Consider two-slit interference for electrons again. According to the many worlds theory, there are no collapses and no hidden variables, so the system is described completely in terms of evolution of the wave function according to Schrödinger's equation. The wave function for the electron splits into two packets, one passing through the left slit and the other passing through the right slit, and beyond the slits the two terms come together and interact to produce the characteristic interference wave pattern at the screen. This wave pattern can be regarded as a superposition of a large number of terms, one for each atom in the screen which might be struck by the electron. Each term in this superposition produces a corresponding term in the later state in which a photon is emitted from that atom, and a corresponding term in the still later state in which the observer sees a flash at that location on the screen. So in this case we have a large number of worlds, one for each location at which the electron might be detected, and in each world the observer sees a flash at a well-defined location.

However, note that even though the electron strikes the screen at a unique location in each world, it does not pass through a unique slit in each world. The term in which the electron strikes the screen at some particular location is causally influenced by both the right-slit wavepacket and the left-slit wavepacket, since the interference between them is what produces the amplitude of the term at the screen. So if the boundaries of worlds are defined by causal isolation, none of the worlds resulting from this interference experiment is one in which the electron passes through a unique slit.

In an important sense, of course, there is just *one* physical world—the one described by the wave function. So in what sense can there also be *many* worlds? DeWitt, who perhaps did the most to publicize Everett's theory, arguably saw the worlds of Everett's theory as an ontological addition to the standard theory—a structure imposed on the wave function from outside (DeWitt, 1970, 33). But this just raises the measurement problem anew: Which physical processes cause one world to split into many, and why (Barrett 1999, 150)? Recently, however, a consensus has arisen according to which the worlds are emergent: They are patterns in the underlying wave function structure of the (unique) physical universe (Wallace, 2010, 2012). When these patterns become sufficiently autonomous—when they no longer interact to an appreciable extent—then for all practical purposes they describe independent worlds with their own individual properties and causal development.

As noted in Chapter 2 (and explored further in Chapter 5), the many-worlds theory does not exhibit the problematic nonlocality of Bohm's theory or the GRW theory. However, it shares the problem of dimensionality with these theories, since in all three cases the quantum state inhabits a high-dimensional space, not the three-dimensional space we are familiar with. Again, this is taken up in Chapter 7. At the quantum level of description, the many-worlds theory is the simplest of the three theories: There are no additional physical processes (such as the GRW theory postulates), and no additional physical entities (such as Bohm's theory postulates), just a wave function evolving according to the Schrödinger dynamics. As in the GRW theory, then, one might say that there are no particles in the many-worlds theory, just a wave that sometimes behaves (within a particular branch) in a particle-like way.

Although the many-worlds theory is simple at the fundamental level, at the level of everyday objects it is strikingly complicated. In the earlier interference experiment, the preexperiment world splits into many postexperiment worlds. Where there was initially a single screen and a single observer, there is now a multiplicity of distinct screens and distinct observers. This naturally raises a host of metaphysical issues concerning the identity of objects and persons. If the many-worlds theory is true, then Parfit's (1971) examples about personal fission are not just thought experiments! It also raises tricky questions concerning probability. According to the many-worlds theory, every outcome of a measurement actually occurs. But then in what sense can we say that some outcomes are more probable than others? How can we relate the probabilities of outcomes to their squared amplitude, as required by the Born rule? These questions are explored in Chapter 6.

Everett's presentation of his theory is somewhat cryptic: He describes his approach as a "relative state" approach, and never mentions the multiplicity of worlds.[9] This has led to a good deal of speculation concerning the nature of the multiplicity at the heart of his approach: Although a multiplicity of worlds (or branches of reality) is the standard way to talk about Everett's theory, others talk in terms of a multiplicity of minds (Lockwood, 1989) or of facts (Saunders, 1995). However, since Everett's theory is simply standard quantum theory without the measurement postulate, and since none of these proposals differs from Everett in this respect, they are all essentially the same theory (Wallace, 2012, 38). Unlike spontaneous collapse theories and hidden variable theories, which can be formulated in various physically distinct ways, there is just one many-worlds theory, although there is some disagreement about how best to express it. In the rest of the book, I take Wallace's version of the many-worlds theory as canonical and set aside the question the extent to which it is faithful to Everett's intent.

3.5 Reducing the Alternatives

I have briefly introduced the three most prominent solutions to the measurement problem: the GRW theory, Bohm's theory, and the many-worlds theory. They clearly offer very different descriptions of the quantum world. And as noted earlier, each of the three is the starting point of a research program. By now there are a number of spontaneous collapse theories and hidden variable theories, differing significantly in their descriptions of the world from the original theory of each kind. Even in the case of the many-worlds theory, where there is just one physical theory, there is some debate about how it describes the physical world. So we seem to have a genuine case of metaphysical underdetermination in quantum mechanics. Furthermore, these are not philosophers' cooked-up alternatives to some scientific theory, but real scientific theories in their own right—serious proposals concerning how the quantum world might be. The project of quantum ontology looks like it boils down to deciding between these competing descriptions. And the worry is that at the end of the day there may be no good way to decide between them, so the project of basing our ontology on our best physical theory will prove to be impossible, at least at this level of description.

However, there is another line of thought running through the history of quantum mechanics according to which there is no genuine underdetermination here. The measurement postulate has always sat uncomfortably with the rest of quantum mechanics. Von Neumann (1932, 186) first explicitly formulated the measurement postulate as part of quantum mechanics, but he also claims that when the measurement postulate applies is arbitrary to a large extent, suggesting that it is not an objective physical process, but something more closely connected to our situation in the world as observers (1932, 224). In that case, maybe the measurement postulate should never have been regarded as part of the physical theory itself; perhaps we should read the textbooks as saying that a system undergoing a measurement *looks as if* its state collapses to an eigenstate.

On that way of construing standard quantum mechanics, the many-worlds theory seems much more natural than Bohm or GRW. Everett can be regarded as simply showing us a consistent way to interpret the standard theory that physicists use, while Bohm and GRW construct alternatives to that theory. But why bother proposing alternatives to the standard theory, given that it is so successful in practice? A natural suspicion is that these alternatives are motivated primarily by "classical prejudice" (Zeh, 1999, 197)—by a stubborn unwillingness to let go certain aspects of our classical worldview. The suggestion is that clear-eyed attention to what standard quantum mechanics actually says shows that there is just one theory of quantum mechanics, and one consistent way of understanding it. As DeWitt famously puts it, "the mathematical formalism of the quantum theory is capable of yielding its own interpretation," and that interpretation is the many-worlds theory (1970, 33).

A stronger argument is sometimes made here as well: that Bohm's theory and the GRW theory aren't just unmotivated as alternatives to the many-worlds theory, but strictly speaking they aren't *alternatives* to it at all. The claim is that Bohm's theory and the GRW theory reduce to the many-worlds theory, so we don't have a genuine example of underdetermination, but just the appearance of underdetermination.

This argument is most frequently made concerning Bohm's theory. The basic insight is that since neither Bohm's theory nor the many-worlds theory involves a collapse process, the wave function evolves over time in exactly the same way according to both theories. Furthermore, both theories take the wave function to describe the evolution of a physical entity in the world.[10] According to the many-worlds theory, everyday physical objects are patterns in this wavefunction-stuff, and when the wave function describing an object branches into multiple terms, the object itself branches into multiple copies. But since exactly the same patterns in wavefunction-stuff are present according to Bohm's theory, Bohm's theory too describes objects branching into multiple copies.

Consider, for example Schrödinger's cat. At the end of the experiment, the quantum state describing the cat has two terms, one of which exhibits a live-cat pattern and the other of which exhibits a dead-cat pattern. According to the many-worlds theory, this pattern in the wavefunction-stuff means that there are two worlds at the end of the experiment, one containing a live cat and the other containing a dead cat. According to Bohm's theory, the wavefunction-stuff exhibits exactly the same patterns, so it looks like there are two worlds according to Bohm's theory too, one containing a live cat and the other containing a dead cat. Bohm's theory also postulates the existence of a set of particles inhabiting one but not the other of these worlds, but the important claim is that the presence of the particles in one world doesn't detract from the reality of the other world. Hence "Bohm's theory contains the *same* 'many worlds' of dynamically separate branches as the Everett interpretation" (Zeh, 1999, 200). As Deutsch puts it, Bohm's theory is just the many-worlds theory "in a state of chronic denial" (1996, 225).

A similar argument has been made concerning the GRW theory (Cordero, 1999). The GRW theory, of course, *does* have a collapse process, so the wave function does not evolve in exactly the same way according to the many-worlds theory and the GRW theory. But nevertheless, it looks like the same patterns in the wave function are present. In the case of Schrödinger's cat, the superposition of a live-cat term and a dead-cat term is unstable: It is rapidly subject to a GRW collapse that multiplies the state by a narrow Gaussian in the coordinates of one particle in the cat. This has the effect of concentrating almost all the wave function amplitude in one term or the other—but not *precisely* all the amplitude. The other term in the superposition remains, albeit with very small amplitude. So neither pattern in the wave function is completely eliminated: one is a high-amplitude pattern and the other is a low-amplitude pattern, but both patterns are (apparently) still there.[11] Since the amplitude of the pattern has nothing to do with the presence or absence of a cat, the postcollapse state is still one in which there is a dead cat (in one world) and a live cat (in another). Hence, Cordero concludes that there is "a Many-Worlds ontology at the heart of theories like... GRW" (1999, S67).

The basic ontological view on which both arguments depend is a kind of functionalism that Wallace calls *Dennett's criterion*: "A macro-object is a pattern, and the existence of a pattern as a real thing depends on the usefulness—in particular, the explanatory power and predictive reliability—of theories which admit that pattern in their ontology" (2003a, 93). Since all three theories—many worlds, Bohm, and GRW—admit the same branching patterns in the wave function, and those branching patterns are crucial to the explanatory and predictive power of the theories, then both the live and dead copies of Schrödinger's cat exist according to all three theories. In particular, the substance in which the pattern is instantiated is irrelevant. One cannot claim that only objects made out of particles (rather than wavefunction-stuff) exist, or that only objects made out of high-amplitude wavefunction-stuff exist, without violating Dennett's criterion.

The goal of both Bohm's theory and the GRW theory is to make sure that an experiment has a single outcome. The import of this line of argument is that both theories fail to achieve their goal, since every possible measurement outcome actually occurs despite the addition of hidden variables or collapses. The only sense in which these theories produce measurement outcomes, then, is the world-relative sense. If this is correct, then the extra mechanisms of Bohm's theory and the GRW theory are doubly redundant: Not only can one do without either mechanism and still solve the measurement problem by adopting the theory, but adding the extra mechanisms doesn't even produce a distinct solution to the measurement problem from many worlds.

This is surprising. After all, at first glance it looks like it is the many-worlds theory that has the redundant structure: All our experience is of a single world, and yet the many-worlds theory asks us to accept the reality of a multitude of worlds. As Bell puts it, "I do not myself see that anything useful is achieved by the assumed existence of the other branches of which I am not aware" (1976b, 16). But a strong case can be made that the other branches of reality are not redundant. Even though physical interaction between the worlds is too small to be directly observable—and indeed that is why the branches qualify as "worlds"—the physical interaction between worlds does not disappear altogether. Since the worlds continually "jostle" each other, and this jostling is crucial to the empirical adequacy of quantum mechanics, the worlds must all be equally real (Deutsch, 1996, 225). Even putting aside the jostling between branches, it looks like the wave function has to be real in order to "steer" the particles (Bohm's theory) or generate the probability distribution for collapses (GRW theory). In the context of Bohm's theory, Bell insists that "no one can understand this theory until he is willing to think of [the wave function] as a real objective field" (1981, 625), and a similar sentiment could be expressed regarding the GRW theory, since in both cases it looks like the wave function has a genuine guiding role.[12] But if the wave function is a real objective field, then the functionalist argument has teeth, since it is hard to deny that a cat-shaped pattern in physically real stuff is a cat.

However, this is not a foregone conclusion.[13] Note that cat-like patterns (whether alive or dead) are exhibited over time, not at a single time: Being a cat (either alive or dead) is a matter of dynamical behavior. But it is far from clear whether we can locate cat-like patterns in the branching structure of the wave function. Cats, as we know them, don't branch into multiple copies. So in this sense, the dynamical pattern exhibited by the branching wave function is very different from the pattern exhibited by the Bohmian particles, since the latter does not branch. So it does not follow that if there is a live-cat pattern in a particular Bohmian state, then there is also a dead-cat pattern, since the wave function in the dead-cat term does not exhibit the same kind of nonbranching pattern that the particles in the live-cat term exhibit. Similarly in the GRW theory, the pattern exhibited by the branching structure overall is very different from the pattern you see if you only follow the large branch, in that there is no subsequent branching in the latter. If there is a live-cat pattern in the large branch, it doesn't follow that there is a dead-cat pattern in the small branch, since the small branch is forever branching into still smaller branches.

An advocate of many-worlds might respond that cats *do* branch into multiple copies: If the many-worlds theory is right, then the cats, like any other objects, are branching entities. But this is not a good strategy to pursue in the current context, because it amounts to conceding that many-worlds cat-patterns are very different from Bohm/GRW cat-patterns: If the many-worlds branching entities are cats, then the Bohm/GRW nonbranching entities are not (and vice versa). Rather, what the many-worlder needs to do to make the case is to find a way of identifying nonbranching cat patterns in the branching wave function structure.

The way to do this is to adapt a method devised by David Lewis. Instead of regarding the branching wave function as describing a branching object, regard it as describing a number of distinct objects whose histories initially coincide but later diverge.[14] Seen in this way, the wave function in the Schrödinger's cat case doesn't represent a single branching entity, but rather multiple nonbranching entities. At any give time, there are multiple cats present; initially they all coincide, but later they diverge into a set of live cats and a set of dead cats, and later each set diverges further. That is, a cat is identified with a *path* through the branching structure of the wave function. The Bohmian particles trace out one such path, as does the large branch in the GRW theory. But there are cat-like paths in the other branches of the wave function, too.

However, it doesn't immediately follow that Bohm and GRW reduce to many-worlds. Recall that *every* physically possible process appears as a branch in the many-worlds wave function. So although we can identify patterns that behave like cats in the wave function, there are also patterns that initially behave like cats, but later start behaving in distinctly noncatlike ways, for example, by spontaneously becoming a gas. Bohm's theory provides us with a rationale for ignoring these paths, namely that they are associated with very small squared wave function amplitude, so it is very unlikely that the Bohmian particles will follow them. Similarly in the GRW theory, it is very unlikely that the large branch will follow such paths, for the same reason. In each case, the additional dynamical law of the relevant theory entails that the probability of noncatlike behavior is low. But the many-worlds theory has no additional dynamics, so initially at least, there is no warrant for ignoring the deviant paths. If we can't ignore them, then we have no reason to expect a pattern that appears cat-like right now to go on behaving in a cat-like way. While cat-like patterns may exist as paths through the wave function, we have no reason to expect the future to follow such a path.

Of course, this is tantamount to saying that the many-worlds theory is empirically inadequate as an account of cats—or indeed of any kind of object. Without a mechanism for identifying squared wave function amplitude with probability, the theory simply cannot reproduce the predictions of the Born rule. If the many-worlds theory is empirically inadequate, then Bohm's theory and the GRW theory do not reduce to it, as they make the correct probabilistic predictions about cats and many-worlds theory does not (Callender 2010, 2015). And by the same token, if the many-worlds theory is empirically inadequate, then the additional mechanisms postulated by Bohm and GRW are not idle modifications to a perfectly good theory, but necessary complications designed to capture the probabilistic facts.

As mentioned earlier in this chapter, there are arguments available according to which the identification of squared wave function amplitude with probability *can* be explained within the many-worlds theory. If those arguments succeed, then the many-worlder can offer the same reason for ignoring low-amplitude paths as the Bohmian and the GRW theorist. The point here is just that these arguments remain controversial, and if they fail, then so does the many-worlds theory. If the many-worlds theory fails in this way, then the empty wave function branches in Bohm's theory and the small wave function branches in the GRW theory don't contain additional copies of the objects described by the Bohmian particles or by the large GRW term, because unadorned wave function branches don't have the requisite probabilistic structure.

So even though the many-worlds theory is simpler than the alternatives, it is not at all clear whether the additional structure postulated by Bohm's theory and the GRW theory is unmotivated, or whether the latter two theories reduce to the former. But even if the argument against underdetermination is not decisive, a conditional version is much more defensible: *If* the many-worlds theory makes the same probabilistic predictions as Bohm and GRW, then the latter essentially reduce to the former, and we really only have one description of the quantum world. If all three theories are empirically adequate, this gives us a precise sense in which the simpler theory wins out.

3.6 Conclusion

Where does all this leave the question of metaphysical underdetermination in quantum mechanics? Although we have three serious contenders for our best theory of the quantum world, and several variants of those, it is not clear that any of them is empirically adequate. The many-worlds theory, as just mentioned, has trouble accounting for probabilities; this issue will be explored in detail in Chapter 6. Bohm's theory and the GRW theory have trouble with locality: It looks like they require instantaneous causation at a distance, and this is hard to square with relativity, as we shall see in Chapter 5. Furthermore, there are experiments that could in principle distinguish between collapse theories like GRW and no-collapse theories like Bohm and many-worlds: If we could determine experimentally whether macroscopic objects remain in superpositions of distinct position states, this would decide between them. Such experiments are not technically feasible at present, but may become so.[15]

If the world is nice to us, we will be left at the end of the day with one empirically adequate quantum theory, and the real underdetermination considered in this chapter will not arise. We can look to this theory to inform us about physical ontology on the quantum scale. But there is no guarantee that the world will be nice to us. It is entirely conceivable that none of the three theories will turn out to be defensible, in which case we will have no guidance at all concerning quantum ontology unless and until some ingenious physicist produces a new theory that solves these difficulties.

Of course, it is also conceivable that more than one quantum theory will prove defensible. Given the argument of the previous section, one might think that the resulting underdetermination could involve at most two theories, and that they would have to be Bohm and GRW (since if many worlds remains a contender, then the other theories reduce to it). But this assumes that the debates over the empirical adequacy of the many-worlds theory (and the others) can be resolved, and one might quite reasonably be less than optimistic about the possibility of definitive philosophical progress. Epistemically speaking, unless the issues surrounding the empirical adequacy of the various contenders are resolved, we still have (at least) three potential descriptions of the microworld on the table.

So does that render hopeless the project of basing our metaphysics of the microphysical realm on quantum mechanics? I don't think so. Note that we have at most a small handful of options. According to the traditional underdetermination worry, since we can presumably concoct an empirically adequate theory with more or less any underlying ontology, science can't tell us anything about metaphysics. But the distinctive underdetermination worry in quantum mechanics stays closer to scientific practice, and it doesn't generate so many alternatives. We may not be able to say definitively what the world is like, but at least we can provide a small range of concrete possibilities.

As far as the project of this book goes, that leaves us with two tasks. The first is to assess the empirical adequacy of the various theoretical options canvassed in this chapter, as discussed earlier. The second is to spell out the ontological consequences of the three main options and their variants. Since the three main options often disagree on ontological matters, it looks like most of the consequences of quantum mechanics for ontology will have to be expressed conditionally—for example, if Bohm's theory is true, then the world is deterministic (at the relevant level of description). But given that *all* the theories under consideration are metaphysically revisionary in some way or another (as we shall see), this is still a worthwhile project. Quantum mechanics is ontologically revolutionary, even if we can't say exactly what form the revolution takes.

4| Indeterminacy

IN CHAPTER 2, WE CONSIDERED THE possibility that quantum systems may simply lack well-defined premeasurement values for certain properties: Although we can measure the z-spin of an electron and get an outcome, perhaps the measurement doesn't reveal a preexisting z-spin property. One way to describe the situation is in terms of indeterminacy: Prior to the measurement (at least), the electron simply lacks a determinate z-spin. Indeed, quantum mechanics is often taken to entail that there is indeterminacy in the world. If this is so, then we can pinpoint at least one sense in which quantum mechanics challenges our classical metaphysical intuitions, since ordinarily we think that the physical world is fully determinate in every detail (although not everyone shares this intuition even in the classical case, as we shall see).

However, quantum mechanics is less than decisive here. (This is a refrain that may become tiresome.) Although indeterminacy fits very naturally within quantum mechanics, it can also be resisted. Nevertheless, quantum mechanics injects new arguments and new considerations into the debate. And more important, perhaps, the kind of indeterminacy that one might be led to posit on the basis of quantum mechanics is very different from the classical kind. To that extent, at least, quantum mechanics tells us something very interesting about indeterminacy.

What do I mean by indeterminacy here? We can draw a useful distinction between determinable properties and determinate properties (Johnson, 1921, 174). The distinction is (like everything) somewhat contested, but the basic idea is straightforward: The determinable specifies a certain range of properties, and the determinate specifies a specific property from within that range. So, for example, color is a determinable property, and specific shades are determinate properties. Coffee mugs have the determinable property color, whereas numbers do not. The coffee mug in front of me is chartreuse; this is the determinate value for the determinable color. Since numbers lack the determinable color, the question of their determinate color doesn't arise. Indeterminacy is the situation in which an object has a determinable property, but no determinate value for that determinable.

A natural view is that all properties in the world are fully determinate: "A physical object is determinate in all respects, it has a perfectly precise colour, temperature, size, etc. It makes no sense to say that a physical object is light-blue in colour, but is no definite shade of light blue" (Armstrong, 1961, 59). In the case of color, this certainly seems plausible. However, for other determinables the situation is less clear. Take size, for example. It is natural to think that my coffee mug has a precise size—a precise determinate value of the size determinable. But what about a cloud in the sky: Does it have a precise size? What about the Australian outback? The existence of objects like this might suggest that there are things that have the determinable size and yet lack a determinate size. In fact, thinking about such cases might convince you that no composite object has a completely determinate size, since it is always indeterminate precisely which smaller objects compose it. Even in the case of my coffee mug, it is plausible that its size is indeterminate at sufficiently high resolution, since for atoms at its boundary it is indeterminate whether they are part of the mug or part of its environment. Indeed, Tye takes compositional vagueness to be necessary and sufficient for there to be objects with fuzzy boundaries: There are objects of indeterminate size if and only if there are objects for which it is indeterminate whether some smaller object is part of it (2000, 201). Since he thinks that compositional vagueness is endemic, Tye concludes that the world is full of objects of indeterminate size (2000, 208).

However, such a conclusion is not forced on us. David Lewis, for example, insists that "the reason it's vague where the outback begins is not that there's this thing, the outback, with imprecise borders; rather there are many things, with different borders, and nobody has been fool enough to try to enforce a choice of one of them as the official referent of the word 'outback'" (1986a, 212). According to this view, the indeterminacy in the size of the outback is entirely linguistic: Physical objects are all perfectly determinate in size, but it is indeterminate which precise collection of them our term "outback" refers to. The dispute here is in large part semantic: Lewis takes it to be indeterminate which precise collection of atoms the term "Mount Everest" refers to, whereas Tye takes it as obvious that "Mount Everest" refers determinately to a single Himalayan mountain (Tye 2000, 196).

In what follows I will assume that Tye is right about this semantic matter. However, my intention here is not to weigh in on this dispute; if you like, substitute Lewis's position for Tye's wherever it occurs. My real intention is to explore the supposition on which the Tye-Lewis debate is based, namely that indeterminacy has to do with composition. Both Lewis and Tye assume that ordinary macroscopic objects are "swarms of particles" (Lewis, 1993, 23; Tye, 2000, 196), and that it is for such swarms that indeterminacy arises. But what of the particles themselves? Given their respective analyses of indeterminacy, Lewis and Tye must both say that the particles themselves cannot be of indeterminate size. It cannot be indeterminate which objects constitute an electron if an electron has no proper parts, so an electron cannot be of indeterminate size on Tye's view. Similarly, it cannot be indeterminate which collection of objects the phrase "this electron" refers to if an electron has no proper parts, so an electron cannot be of indeterminate size on Lewis's view. This may not bother them unduly; given a classical, Newtonian understanding of fundamental ontology, there is no indeterminacy at the level of individual particles, so no such analysis is necessary.

But of course the Newtonian understanding of the world has been superseded by quantum mechanics, so it is worth inquiring anew whether there is indeterminacy in the world. As Sanford (2011) notes, this seems to be a question for physics: "Logic and metaphysics cannot answer this question from its own resources. Science textbooks represent particles and atoms as clouds. Textbook writers fifty years ago knew that the picture of perfect little spheres, the electrons, in elliptical orbits around a nucleus was misleading. Now the picture is simply obsolete." That is, Sanford takes one of the lessons of the transition from classical physics to quantum physics to be that particles can be cloud-like, and hence of indeterminate size. And if this is so, then both Lewis's and Tye's analyses of indeterminacy are at best incomplete, since they do not countenance indeterminacy at level of individual particles.

However, we need to proceed carefully here. It is true that when discussing quantum mechanics, physics textbooks often represent particles in a cloud-like way; for example, Figure 4.1 shows a typical diagram of an electron orbiting an atomic nucleus. But even if quantum mechanics represents electrons as cloud-like, it doesn't follow that physical reality is indeterminate. As every reputable textbook is quick to point out, what is actually represented in diagrams like Figure 4.1 is the probability density for finding the electron in a particular location, given according to the Born rule by the square of the wave function amplitude. That is, the cloud-like picture is a representation of the epistemic facts about the electron, and it may or

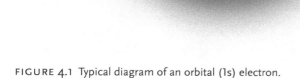

FIGURE 4.1 Typical diagram of an orbital (1s) electron.

may not correspond to an underlying ontology that is also cloud-like. It is true that the wave function itself in this situation looks rather like a cloud around the nucleus. But as we saw in Chapter 3, there are several competing versions of quantum mechanics that give different interpretations to the wave function.

The task of this chapter is to investigate what quantum mechanics actually tells us about the existence of indeterminacy of the physical world. The results, as we shall see, are somewhat equivocal: Both the existence and the extent of quantum indeterminacy are matters of debate and interpretation. But insofar as quantum mechanics does posit indeterminacy, that indeterminacy has nothing to do with composition or even with familiar kinds of vagueness, and hence quantum mechanics changes the nature of the debate over indeterminacy.

4.1 Textbook Indeterminacy

Recall from Chapter 1 the textbook position on the relationship between quantum states and physical properties, namely that each distinct physical property of a system is represented by an eigenstate of the corresponding operator. In matrix mechanics, these eigenstates are mutually perpendicular vectors. For example, the spin of an electron along some particular direction—say the z-axis—can take two values, and hence is modeled by two mutually perpendicular vectors. One vector represents the electron as z-spin-up, the other represents the electron as being z-spin-down, and those are the only two z-spin properties the electron can have. The problem that has much concerned us over the course of the previous chapters is that there is a continuum of states between the eigenstates; if the electron has exactly two possible z-spin properties, then what do all the vectors that lie between the two mutually perpendicular ones represent? The prospect of indeterminacy in the world provides us with a response to this problem, namely that the in-between states represent the system as lacking a determinate value of the determinable property represented by the operator. This is a fairly standard way of understanding quantum states, commonly known as the *eigenstate-eigenvalue link*, or (since that is a bit of a mouthful) the *strict link*:

> **Strict link:** A system has a determinate value for a given determinable property if and only if its state is an eigenstate of the operator corresponding to the property, and the determinate value is the eigenvalue for that eigenstate.

So for the electron, all the states in the vector space represent the electron as having the determinable property z-spin, but only two of those states represent the electron as having a determinate value for that property, namely the two eigenstates. For the rest of the states, the z-spin of the electron is simply indeterminate.

The strict link is often explicitly appealed to by philosophers (as we shall see), but it is also implicit in the treatment of physical properties in many physics textbooks. For example, Baym considers the operator that represents angular momentum around the z axis and notes that "photons that are in an eigenstate, $|R\rangle$ or $|L\rangle$, of this operator can be assigned a definite value of the z component of angular momentum ... Any other photon state $|\psi\rangle$ cannot be assigned a definite value of angular momentum" (1969, 17). Eigenstates are states in which systems have definite or determinate values for the determinable property concerned, and noneigenstates are states in which they lack determinate values. Hence, one might reasonably infer from reading physics textbooks that quantum mechanics postulates indeterminacy in the physical world.

It is worth noting at the outset that this kind of indeterminacy is quite markedly different from the indeterminacy countenanced by Tye, since it has nothing to do with composition. As far as we know, electrons have no proper parts, and yet according to the strict link electrons can exhibit indeterminacy. Even if, unbeknownst to us, electrons are composed of parts, the indeterminacy proposed here makes no reference to any such parts. Nor is there vagueness here, at least of any familiar kind. Familiar vagueness involves a continuum of related properties, such as the range of colors between red and blue, and indeterminacy arises because there is no sharp boundary between red and purple. It is tempting to think of the cloud-like electron of Figure 4.1 in these terms: It has an indeterminate boundary because there is a continuum of properties—the density of the cloud—and no sharp boundary between the cloud and empty space. But consider an electron e whose state is

$$\frac{1}{\sqrt{2}}(|\uparrow\rangle_e + |\downarrow\rangle_e), \qquad (4.1)$$

where $|\uparrow\rangle_e$ and $|\downarrow\rangle_e$ are eigenstates of z-spin. The electron's spin is indeterminate, but not because it is indeterminate where in a continuous range of spin properties the boundary of "spin-up" should be; there are just two discrete determinate spin properties that the electron may have or lack.[1] Of course, even though the spin properties of an electron are discrete, there is a continuous range of quantum states between the eigenstates. But there is no real vagueness here either, at least according to the strict link. The boundaries between the states in which the electron has the spin-up property, the states in which it has the spin-down property, and the states in which it simply lacks a spin property are perfectly precise according to the strict link.[2]

Indeterminacy in the world based on composition or vagueness (or both) may be philosophically controversial, but at least it is relatively familiar: When Tye says that the spatial extent of Mount Everest is indeterminate, we may not agree, but we know what he means. The proposal embodied in the strict link is more radical and unfamiliar; it is just a basic, unanalyzable fact about determinable properties that they can have or lack determinate values. Let us call this novel kind of indeterminacy *quantum indeterminacy*.

On this understanding of quantum mechanics, indeterminacy is endemic in the physical world, and the indeterminacy is unanalyzable, having nothing to do with composition or vagueness. But this isn't necessarily problematic. Unanalyzable indeterminacy may be conceptually alien, but there is no obvious incoherence in the idea. Barnes and Williams (2011) defend an account of metaphysical indeterminacy according to which metaphysical indeterminacy is *brute*; it is an irreducible aspect of the world. Nevertheless, they contend that we can understand what metaphysical indeterminacy is via our pretheoretic notion of indefiniteness. To say that something is metaphysically indeterminate is just to say that it is indefinite, where the source of the indefiniteness is the nonrepresentational world rather than our representations. This seems to be just what we need as an account of quantum indeterminacy. What's more, this understanding of quantum states seems to show a healthy respect for the primacy of empirical science: If our best-confirmed theory of the physical world, taken literally, represents electrons as lacking determinate positions, then we should accept that there is indeterminacy in the world, rather than claiming that the theory is incomplete just to preserve our classical intuitions.

But this sanguine attitude only seems appropriate if we can we keep quantum indeterminacy confined to the microscopic world. Maybe we can convince ourselves that we sometimes see composite physical objects of indeterminate spatial extent, for example when we see Mount Everest. But surely we never see physical objects with no determinate position whatsoever! If quantum indeterminacy is a feature of the physical world, why don't we ever experience it? The standard textbook response at this point is to appeal to the measurement postulate: Baym (1969) does exactly that in the sentence following the earlier quote. The measurement postulate, recall, tells us that if we measure a determinable property of a system, the probability of getting a particular determinate value for that property is given by the square of the projection of the state onto the corresponding eigenstate.[3] So even though the electron in state (4.1) does not have a determinate spin, when we measure its spin, we get the results spin-up and spin-down with probabilities 1/2 each. Furthermore, the measurement process results in the electron's state "collapsing" to the eigenstate corresponding to the measurement outcome, a state in which it does have a determinate spin. This ensures that we never encounter any indeterminacy; physical systems can have indeterminate properties, but this indeterminacy conveniently disappears whenever we observe the system in question.

So the measurement postulate succeeds at keeping quantum indeterminacy where we can't directly experience it. But as we saw in Chapter 3, the measurement postulate is indefensible, and different versions of quantum mechanics replace it in different ways. Later in this chapter, we will look at the extent of quantum indeterminacy according to each of these versions. But first, let us consider a more radical proposal:

Perhaps the existence of quantum indeterminacy renders these different versions of quantum mechanics unnecessary in the first place. One way of understanding the role of these alternatives is to rescue determinacy at the macroscopic level—to ensure that pointers on measuring devices always have determinate positions. But perhaps the world is more indeterminate than we take it to be; perhaps pointers on measuring devices *don't* have determinate positions after all. This might seem like a crazy idea—and it is—but as we shall see, it is surprisingly hard to refute.

4.2 Radical Indeterminacy

Recall from Chapter 3 what happens when we perform a z-spin measurement on an electron in state (4.1). If we reject the measurement postulate, the result is

$$\frac{1}{\sqrt{2}}(|+\rangle_o|+\rangle_s|+,\uparrow\rangle_e + |-\rangle_o|-\rangle_s|-,\downarrow\rangle_e), \tag{4.2}$$

where $|+\rangle_o|+\rangle_s|+,\uparrow\rangle_e$ is an eigenstate of the electron being spin-up and moving upward, of a flash appearing in the upper half of the screen, and of the observer seeing a flash in the upper half, and $|-\rangle_o|-\rangle_s|-,\downarrow\rangle_e$ is an eigenstate of the electron being spin-down and moving downward, of a flash appearing in the lower half of the screen, and of the observer seeing a flash in the lower half. In our discussion of the many-worlds theory, we considered two options for interpreting this state. One is to claim that (4.2) is *both* a state in which the electron hits the upper part of the screen *and* a state in which the electron hits the lower part of the screen. This is the option pursued by the many-worlds theory. But we also set aside another option, namely that (4.2) is *neither* a state in which the electron hits the upper part of the screen *nor* a state in which the electron hits the lower part of the screen.[4] This option, which follows from the strict link, takes the outcome of the experiment to be *indeterminate*.

This does not look like a promising direction for exploration. Perhaps there is indeterminacy in the world, but surely the observable outcome of an experiment isn't indeterminate—unless we've designed the experiment very poorly indeed! We can countenance indeterminacy at the microscopic level: Perhaps an electron's spin or location can be indeterminate, since we never directly observe the properties of an electron. Even at the macroscopic level, certain kinds of indeterminacy might be possible in cases of vagueness. But while vagueness might make the size of a flash on a screen somewhat indeterminate, surely it can't be indeterminate whether the single flash appears in the middle of the upper half of the screen or in the middle of the lower half.

However, the possibility of macroscopic indeterminacy of this kind can be defended. The basic argument is that quantum mechanics itself prevents this indeterminacy from being noticed by us. Note that according to the strict link, (4.2) is not a state in which the observer has a determinate experience concerning the location of the flash. This introduces a sense in which the indeterminacy in the location of the flash is not detectable by the observer. You might reasonably argue that in order for an observer to be able to reliably detect some outcome, observing the system in question must produce one determinate experience if the outcome occurs and a different determinate experience if the outcome doesn't occur. If that is the requirement, then the observer cannot detect the indeterminacy in the location of the flash, because the indeterminate state of the screen does not produce a determinate experience in the observer. Without a determinate experience, the observer is in no position to identify the state of the screen as indeterminate. So given that human observers are basically quantum mechanical systems, then we have the beginnings of an argument that even nonvague macroscopic indeterminacy would be undetectable by us.

But it is only the beginnings of an argument. The obvious rejoinder is that we know directly that our own experiences are determinate. Observers in such experiments report determinate experiences, either a flash in the top half of the screen or a flash in the bottom half of the screen, and if you don't believe their reports, you can try it yourself. That is, it is not a requirement for detecting an outcome that there are two different determinate experiences, one if the outcome occurs and the other if it doesn't. All you need is a detectable difference in experience, and the difference between a determinate experience and an indeterminate experience is detectable. So macroscopic, nonvague indeterminacy in the world would be detectable by us, and in fact we don't detect it.

So any argument that indeterminacy in the world would be undetectable needs to do better than that, and in fact a better argument has been suggested by Albert (1992, 118) and developed by Barrett (1999, 97). The strategy is to ask the observer whether her experience is determinate and see what she says. Presumably if her experience is determinate, she will answer "Yes." We can model this process quantum mechanically by including the state of the observer's vocal apparatus in our schematic quantum representation. So let $|o\rangle_v$ be the eigenstate of the observer's vocal apparatus in which she has yet to make an utterance, $|Y\rangle_v$ be the eigenstate in which she utters "Yes," and $|N\rangle_v$ be the eigenstate in which she utters

"No." Then we know that if her initial state is $|\text{o}\rangle_v|+\rangle_o$, her experience is determinate, so her state after she is asked whether her experience is determinate becomes $|Y\rangle_v|+\rangle_o$. Similarly, if her state is $|\text{o}\rangle_v|-\rangle_o$, her experience is also determinate, so her state after she is asked whether her experience is determinate becomes $|Y\rangle_v|-\rangle_o$. But in fact the state of the electron-screen-observer system at the end of the experiment is

$$\frac{1}{\sqrt{2}}(|\text{o}\rangle_v|+\rangle_o|+\rangle_s|+,\uparrow\rangle_e + |\text{o}\rangle_v|-\rangle_o|-\rangle_s|-,\downarrow\rangle_e). \tag{4.3}$$

So appealing to the linearity of quantum mechanics (Chapter 3), the state of the system after the observer is asked whether her experience is determinate becomes

$$\frac{1}{\sqrt{2}}(|Y\rangle_v|+\rangle_o|+\rangle_s|+,\uparrow\rangle_e + |Y\rangle_v|-\rangle_o|-\rangle_s|-,\downarrow\rangle_e), \tag{4.4}$$

which can be rewritten

$$|Y\rangle_v \frac{1}{\sqrt{2}}(|+\rangle_o|+\rangle_s|+,\uparrow\rangle_e + |-\rangle_o|-\rangle_s|-,\downarrow\rangle_e), \tag{4.5}$$

since $|Y\rangle_v$ can be factored out of both terms. That is, this state is an eigenstate of the observer uttering "Yes," and hence the observer's utterance is determinate. She says "Yes."

So even though the observer's experience is not determinate, she reports that it is. And presumably we can take this report as a reflection of her belief: She *believes* that her experience is determinate, even though it is not. (Alternatively, we could model the observer's belief about her experience directly, and by an exactly analogous argument come to the conclusion that she believes that her experience is determinate.) So we now have a much stronger argument that indeterminate experiences are undetectable by us: When our experience is indeterminate, we believe that it is determinate.

On this basis, Albert (1992, 124) and Barrett (1999, 94) go on to entertain the possibility that quantum mechanics without the measurement postulate—which they call the *bare theory*—could be a perfectly good, empirically adequate physical theory, distinct from the many-worlds theory. The bare theory, like the many-worlds theory, takes quantum mechanics without the measurement postulate to be a complete physical theory. But unlike the many-worlds theory, the bare theory does not attempt to find a structure of branches or worlds in a state like (4.2), and hence it does not ascribe properties to systems relative to such worlds. Rather, properties are ascribed to a system by looking at the state *as a whole*, and when the state as a whole is not an eigenstate corresponding to the property in question, then it is simply indeterminate whether the system possesses the property or not.

The idea behind the bare theory is that the measurement postulate is simply redundant, because the indeterminacy in the world that it was supposed to eliminate is made undetectable by quantum mechanics itself. This is a very interesting idea. But it is also a very radical idea, as both Albert and Barrett are keen to point out. Indeterminacy does not only arise in esoteric laboratory situations such as measuring the spin of an electron. The quantum dynamics entails that eigenstates are unstable across a wide variety of situations. So, for example, a particle state that is initially an eigenstate of occupying some particular small range of positions will generally evolve over time to a state that is not an eigenstate of this or any other small range of positions. The same goes for the locations of macroscopic objects; a coffee mug whose state is initially an eigenstate of being located on my desk will generally evolve over time to a state that is not an eigenstate of this or any other location. Nor will it remain in an eigenstate of being mug-shaped. So even if by some miracle the state of the world at some time has the familiar determinate properties we typically attribute to it, quantum mechanics ensures that this state of affairs doesn't last for long. If the bare theory is true, indeterminacy is ubiquitous: No object, large or small, has a determinate color, shape, or location. If the bare theory is the right way to understand quantum mechanics, then quantum mechanics entails indeterminacy in the world on a scale that nobody previously imagined.

But can the bare theory possibly be the right way to understand quantum mechanics? Could such widespread indeterminacy at the macroscopic level really go unnoticed? There are several ways you might try to resist the Albert-Barrett argument. You might insist that determinacy is simply *given* in experience—that the determinacy of our experience is a fixed point that must be accommodated by any scientific theory, and hence which cannot be undermined by quantum mechanics. But to insist upon the primacy of the way things seem to us over what science tells us is a very dubious strategy.[5] Along the same lines, one might object that the bare theory fails to explain the nature of our experience, and so fails to be empirically adequate. That is, it does not merely seem to me that I am having *some* determinate experience right now, it seems to be that I am having a particular determinate experience, for example seeing a coffee mug on a desk. This objection, though, tacitly denies the possibility that we are grossly mistaken about our own experience, and hence begs the question against the bare theory.[6]

So I think that Albert and Barrett are correct that the bare theory cannot be challenged by a simple appeal to experience. Nevertheless, neither Albert nor Barrett thinks that the bare theory is acceptable an interpretation of quantum mechanics. Their reasons have to do with the extraordinarily skeptical nature of the bare theory. As Barrett notes, the bare theory's account of experience "makes Descartes's demon and other brain-in-a-vat stories look like wildly optimistic appraisals of our epistemic situation" (1999, 94). But how does this skepticism entail that the bare theory is untenable? Albert (1992, 124) and Barrett (1999, 116) both point out that if the bare theory is true, then we have no *evidence* that it is true. If we believe the bare theory, it is presumably because it is the best way to understand quantum mechanics, and quantum mechanics is highly confirmed. But the evidence that confirms quantum mechanics consists of certain physical events—the outcomes of experiments—that are observed and reported by physicists. According to the bare theory, these experiments didn't produce determinate outcomes, and the physicists didn't have determinate experiences or report determinate results, so there is no evidence for quantum mechanics, and hence no reason to believe the bare theory. Barrett calls a theory that, if true, prevents us from having empirical justification for its truth *empirically incoherent* (1999, 116). He takes empirical incoherence to rule out a theory as an object of scientific knowledge: "such a theory might be true—it is just that if it were true, then one would never know that it was" (Barrett, 1999, 117). Since the bare theory is empirically incoherent, then even though it might be true, it fails as a *scientific theory*.[7]

This seems right: If the bare theory were true, then empirical science would be impossible. So if we are to get on with empirical science, we have to set aside the bare theory. But such practical considerations don't do much to address the skepticism produced by the bare theory. By way of analogy, consider the hypothesis that most sentient beings are brains in vats. Although this hypothesis could be true, presumably we have no empirical evidence for it; I for one have never seen an envatted brain. But suppose I did have the relevant evidence; suppose that my experience suggests that four out of five infants are envatted at birth and remain that way for life. Then it seems I have reason to take the hypothesis seriously. The hypothesis, of course, is empirically incoherent: If it is true, then chances are the experiences that justify it were fed into my envatted brain by a computer, and so tell me nothing about the world. If I want to find out about the world, I have to set aside the brain-in-a-vat hypothesis. But still, I cannot simply ignore the doubt that the hypothesis produces. I can investigate the world if I like, but I can't set aside the worry that the premise on which this investigation is based—that I am not a brain in a vat—is false. Similarly, maybe I should set aside the bare theory if I want to do science, but I can't set aside the worry that any such investigation is pointless.

This is troubling. The bare theory makes epistemic skepticism pressing in a way that it never was before. There is no evidence for an evil demon or for widespread envatment, but there is evidence for the bare theory. This evidence is self-undermining, to be sure, but it is sufficient to generate skepticism, just as seeing envatted brains would generate skepticism. As a practical matter, we might set the skeptical hypothesis aside, but that doesn't make the skepticism go away.

However, there may be other reasons not to take the bare theory seriously. Albert (1992, 124) hints at these reasons when he notes that the bare theory entails that it is not a determinate fact that there are sentient beings: The quantum state of the world is likely to be a superposition of terms only some of which describe sentient beings at all. But of course the bare theory, like any physical theory, could be true even if there are no sentient beings, and we cannot simply insist that we know that there are sentient beings without assuming the kind of epistemic access that the bare theory denies. A more promising line of attack is that the arguments in favor of the bare theory presuppose the existence of sentient beings performing experiments, and yet the bare theory also denies this presupposition. This means that the actual arguments given earlier to defend the bare theory against charges of empirical inadequacy cannot be made (Barrett, 1999, 119). But this does not mean that the bare theory cannot be defended, because arguments without the presupposition might be available.

Here is how such an argument might go. Suppose the state of the world is a complicated superposition of terms, some of which represent me as having a spin-up experience, some of which represent me as having a spin-down experience, and most of which don't represent me (or any other sentient being) at all. Any physical process, introspective or otherwise, that were it instantiated would reliably distinguish between a spin-up experience and a spin-down experience would fail to produce a determinate output in this case. So even though it is not a determinate fact that there are spin-up experiences, the state of the world is not determinately distinguishable from a state in which there are spin-up experiences. Furthermore, any physical process, introspective or otherwise, that were it instantiated would reliably distinguish between sentience and nonsentience will fail to produce a determinate output here, too. So even though it is not a determinate fact that there are sentient observers, the state of the world is not determinately distinguishable from a state in which there are sentient observers. This is (arguably) sufficient to make the bare theory immune from empirical refutation. In general, then, it looks like epistemic arguments against the bare theory are inconclusive. The bare theory entails a radical indeterminacy in our experience and in the world, but since it also entails that indeterminacy is not distinguishable from determinacy, the indeterminacy cannot be used against it.

If there is a conclusive argument against the bare theory, I suspect, its character is semantic rather than epistemic. Consider the *content* of the bare theory—what its terms *mean*. Recall the way that the theory of quantum mechanics is constructed. Determinate states of affairs, such as an electron being spin-up in a certain direction, or a flash appearing in a certain region of a fluorescent screen, are represented by eigenstates. General quantum states are obtained by superposing the eigenstates in various proportions. We struggled earlier with the interpretation of these general states—that is, with what it means when we say that the state of an electron is described by a superposition state like (4.1). The measurement postulate gives us a way of approaching this question; it tells us what experiences we can expect this state to produce under certain circumstances. This may not be a complete or adequate interpretation of the state, but it is a start. But the bare theory, of course, eschews the measurement postulate. So now what does a state description like (4.1) mean?

If the meaning of a state like (4.1) cannot be approached by considering its behavior on measurement, the other possibility is to approach it via the way the state is constructed out of basic states. That is, if we knew what it meant to say that an electron is spin-up in a certain direction or that a flash appears in a certain region of a screen, then we would at least have the beginnings of an interpretation of general superposition states. This interpretation would not be complete or adequate as it stands, but it would be a start. Presumably the meaning of "There is a flash in the top half of the screen" is more straightforward than "The electron is spin-up along z": We think we know what the former means via direct experience, more or less. But the bare theory takes away this approach to meaning, too; according to the bare theory, we never experience flashes on screens. Without any corresponding experiences, an empiricist account of the meaning of "There is a flash in the top half of the screen" is clearly out of the question. By the same token, a Gricean account (Grice, 1957) in terms of speakers' intentions is ruled out, since the bare theory denies that there are speakers with determinate intentions. A Kripkean account (Kripke, 1980) in terms of the history of use of the expression is ruled out, since the bare theory denies that there is any such history. A Davidsonian (Davidson, 1973) account in terms of maximizing the truth of the speaker's total set of utterances is ruled out, since the bare theory denies that there are any such speakers or any such utterances. And so on ad nauseam for any account of meaning one might construct; Whatever features of the world one appeals to to ground meaning, the bare theory denies that the world has those features.[8] Hence, the bare theory is, if true, meaningless. Let us call such a theory *semantically incoherent*. Clearly a semantically incoherent theory cannot be true; if it is true, then it is meaningless, and hence neither true nor false.[9] This, I submit, is the real reason we should reject the bare theory.

The bare theory entails a truly radical indeterminacy in the world. If the bare theory is true, then just about every familiar determinable property is indeterminate. And this is not the familiar indeterminacy due to vague boundaries within continuous ranges of properties; it is not just that an electron orbiting a nucleus or a mug of coffee has fuzzy edges like a cloud, but rather that electrons and mugs and everything else cannot be assigned even an approximate location like "within 50 miles of Miami." As we have seen, this across-the-board indeterminacy is the downfall of the bare theory: It undermines the very meaning of the theory itself. So quantum mechanics cannot entail such thoroughgoing indeterminacy. But this does not rule out indeterminacy of some sort as a consequence of quantum mechanics. In the next section, I look at ways in which quantum indeterminacy of a more moderate extent might be defended.

4.3 Moderate Indeterminacy

The previous section can be viewed as an attempt to solve the measurement problem using only an appeal to indeterminacy—without either new physics or a many-worlds ontology. This noble attempt is (everyone agrees) a failure. So let us return to the solutions to the measurement problem introduced in the previous chapter, and see to what extent they posit indeterminacy in the world. For reasons that will become clear, this topic has received most attention in the case of the GRW theory, so we will start there.

Recall that the GRW theory addresses the measurement problem by postulating a new physical process, whereby each particle has a small probability per unit time of undergoing a "hit" that multiplies its state by a narrow Gaussian centered on a randomly chosen point. Although the state of the particle is (in an obvious sense) highly localized after a hit, its state is not an eigenstate of occupying any particular location, as the Gaussian has tails stretching to infinity. Similar comments apply to macroscopic systems.

In the case of Schrödinger's cat, although the GRW hit process results in a high amplitude for either the live-cat term or the dead-cat term, the other term remains, albeit with very small amplitude.

The strict link makes the presence of these small terms look like a problem. The posthit state of a particle is not a position eigenstate, or even an eigenstate of being located within any finite range. The posthit state of Schrödinger's cat is not an eigenstate of being alive or being dead. And in general, the states of systems will not be eigenstates of any of the properties we ordinarily attribute to them. So according to the strict link, all these properties are indeterminate. If we combine the GRW theory with the strict link, we end up with the radical indeterminacy of the bare theory.

But surely the GRW hit process does *something* to reduce indeterminacy; the posthit states are, after all, very close to eigenstates (on an appropriate measure). The strict link disguises this progress by insisting that only the eigenstate corresponds to possessing a determinate property. So it looks like the strict link is just too strict, and we need something looser. That is, what we need is a link that says that states sufficiently close to eigenstates have determinate properties as well. The *fuzzy link* (Albert & Loewer, 1996, 87; Clifton & Monton, 1999, 699) provides a precise way of doing just that:

> **Fuzzy link:** A system has a determinate value for a given determinable property if and only if the squared projection of its state onto an eigenstate of the corresponding operator is greater than $1 - P$, where the determinate value is the eigenvalue for that eigenstate.

Here P is a parameter expressing what counts as sufficiently close: The smaller P, the closer the state needs to be to the eigenstate to count as having the property. So, for example, a Schrödinger's cat state like $a|\text{alive}\rangle + b|\text{dead}\rangle$ is a state in which the cat is alive provided that $|a|^2 > 1 - P$. The reason for this formulation is to respect the empirical content of the measurement postulate: The squared projection of the state onto the eigenstate is the probability that the system will manifest the corresponding property on measurement. Hence, according to the fuzzy link, there is a probability of more than $1 - P$ that a system that counts as having a given property will behave accordingly, and a probability of less than P that it will behave otherwise.

Prima facie, the fuzzy link isn't fuzzy: There is no more vagueness here than in the strict link, since the fuzzy link, like the strict link, has a sharp cutoff between the states in which the system has a determinate property value and the states in which it lacks one. Nevertheless, it is reasonable to think that vagueness does arise in the fuzzy link because there is no fact of the matter about the location of the boundary P. Presumably the value of P we are willing to endorse is a pragmatic matter rather than a logical or empirical matter: Certainly P must be less than $1/2$, and presumably it should be much closer to zero, but it looks like there is no way to specify a value precisely.[10] In that case, the fuzzy link reintroduces vague boundaries into the quantum world. But note that it is still not vagueness of the familiar kind, because there is no continuum of property values in which this vague boundary occurs.

Suppose we combine the GRW theory with the fuzzy link. What does the resulting theory entail about indeterminacy in the world? For microscopic systems, the GRW hit process is very unlikely to apply, so at this level the indeterminacy examined in the previous section remains: Fundamental particles like electrons typically lack determinate values for their physical properties. This indeterminacy, as remarked earlier, is quite different from that typically countenanced in the philosophical literature on the subject, because it has nothing to do with composition or with the familiar kind of vagueness. But given that it is confined to the unobservable microscopic realm, it is arguably not deeply problematic. The same cannot be said for indeterminacy in the properties of macroscopic objects, as we saw in the previous section. Unrestricted indeterminacy makes for a semantically incoherent theory. But the modified quantum theory we are countenancing here does not have this consequence: The hit process entails that the states of macroscopic objects are generally close enough to eigenstates of ordinary properties like location that they count as having determinate values of those properties according to the fuzzy link. So it looks like quantum indeterminacy has been successfully corralled where it can do no harm.

However, there are at least two ways in which moving from the strict link to the fuzzy link might be considered problematic. First, as noted earlier, the fuzzy link countenances the possibility—albeit with a small probability—that a state in which a system has one determinate property can behave as if it has a different property. You might regard this as a conceptual impossibility, particularly if you equate the possession of a property with the propensity to behave a certain way. For example, you might take it that what it means for an object to be determinately located in a region is that it is certain to be found there. On that criterion, the fuzzy link is ruled out a priori.

Perhaps what Bell (1990, 18) calls "FAPP reasoning" can come to our rescue here. "FAPP" stands for "for all practical purposes," and it is worth stressing that the modification to the Schrödinger dynamics is such that the states of ordinary macroscopic objects will typically be very, very close to eigenstates of the properties we ascribe to them. This means that the probability P appearing in the fuzzy link can be made very, very close to zero, and hence that the chance that an object with one property will behave as if it has a different property is extremely small. For example, according to the GRW theory, there is a nonzero chance that sometime during the next day my coffee mug will behave as if it is inside my desk drawer, even though according to the fuzzy link it is determinately on top of my desk. But this nonzero chance is much less than 1 part in $10^{10^{34}}$ (Bassi & Ghirardi, 1999, 728); we would have to wait much, much longer than the age of the universe to stand a nonnegligible chance of seeing such behavior! So although the theory entails that when the mug is on the desk it sometimes behaves as if it is not, the small value of the probability means that *for all practical purposes* it behaves as it should. That is, on any practical construal of "certainty," you can be certain that the mug is on the desk, even though its state is not precisely an eigenstate for this property.

Perhaps the objector will insist that FAPP reasoning has no place in logic—that the mere possibility of anomalous behavior is enough to mire the fuzzy link in contradiction, even if that behavior is so rare as to be unobservable. If so, there is another strategy available, which is to reformulate the fuzzy link slightly:

Vague link: A system has a determinate value for a given determinable property to the extent that the squared projection of its state onto an eigenstate of the corresponding operator is close to 1, where the determinate value is the eigenvalue for that eigenstate.

The vague link differs from the fuzzy link in that the fuzzy link makes determinate property possession an all-or-nothing thing, whereas according to the vague link determinate property possession is a matter of degree.[11] According to the vague link, for a system to *completely* possess a determinate value for location, its state must be *exactly* an eigenstate of location, and hence the chance of finding it in a different location is zero. This defuses the earlier objection. However, a system can also *mostly* possess a determinate value for location if its state is close to the corresponding eigenstate. In that case, there is a small but nonzero chance of finding it elsewhere—and in fact, you might explain this chance in terms of its *slight* possession of the other determinate property. According to the vague link, my coffee mug almost entirely possesses the determinate property of being on top of my desk, but it also very slightly possesses the determinate property of being inside the drawer. Because the degree of possession of competing properties is so slight, for all practical purposes I can say that the coffee mug is on the desk.

The vague link, as its name implies, explicitly posits vagueness in the world; The boundary of determinate property possession is a vague one. But again, this isn't exactly the familiar kind of vagueness: It is not that there is a continuum of determinate properties, but rather each property comes in degrees of determinacy. This vagueness allows the logician her scruples: A system only *fully* possesses a property if it always behaves consistently with that property. But it does so at the cost of introducing a kind of waffling over determinate property possession: We can't say that macroscopic objects fully possess the properties we ascribe to them. So can we really say that the object *determinately* possesses those properties?

My suspicion is that these kinds of infelicities are inevitable and harmless. We are trying to describe a world whose fundamental nature (according to quantum mechanics) is quite alien to our ordinary conceptions, the conceptions that our language was designed to accommodate. In particular (and somewhat ironically), quantum states are continuous where our ordinary conception of properties requires discreteness. The state of the mug in which it is on the desk is (according to our ordinary conception) discontinuous with the state in which it is in the drawer, but according to quantum mechanics these two states are the extrema of a continuum of intervening states.[12] According to our ordinary conception of the properties of mugs, perhaps it is analytic that if the mug is on the desk, then it is never found in the drawer. But the lesson of quantum mechanics (at least in this formulation) seems to be that we have to give up our ordinary conception and adopt something else instead—something that includes either the fuzzy link or the vague link. Then the analytic consequence of our old conception is moot; it really doesn't matter what it entails, because we have to replace it with something else.

I said earlier that there are two potential problems for the fuzzy link. The first is the one we have just been considering—that the fuzzy link is ruled out by the logic of properties. The second one is that it arguably doesn't keep indeterminacy corralled at the microscopic level after all. Consider the choice of the value of parameter P in the fuzzy link. There is a trade-off inherent in this choice. I mentioned earlier that if P is chosen to be sufficiently close to zero, then an object that counts as having some determinate property almost never behaves otherwise. While macroscopic systems are strongly affected by GRW collapses, microscopic systems are hardly affected at all, and hence the states of microscopic systems are rarely even close to eigenstates of properties like location. But what about in-between systems, like specks of dust? Such systems are affected to an intermediate degree; their states are generally somewhat close to the relevant eigenstates, although they can remain far from eigenstates for appreciable periods of time. But given the right lighting conditions, we can *see* specks of dust, so presumably they should have determinate locations. If we set P far enough from zero that such objects are assigned determinate locations by the fuzzy link even when their states are somewhat distant from the relevant eigenstate, then they will have a nonnegligible probability of behaving in a way that is not consistent with their determinate properties. On the other hand, if we set P close enough to zero that anomalous behavior has a negligible probability, then specks of dust will quite often have no determinate location.

This point can be put more forcefully. It is an assumption of the GRW approach that measurements involve correlating the state of a system with the position of a macroscopic solid object like a pointer on a dial; it is the positions of such objects to which the GRW theory gives well-defined values. But the assumption that macroscopic objects are special can be called into question. A simple way to measure the location of an electron is to run it into a fluorescent screen and observe the resultant flash. There is nothing like a macroscopic pointer here; the production of the flash involves just a few electrons in the screen and a few photons that travel to the observer's eye. As Albert and Vaidman (1989) point out, not enough particles are involved to trigger a GRW collapse, so if the electron is initially in a superposition state in which it cannot be ascribed a position, the photons cannot be ascribed positions either. However we choose the value of P, the GRW theory fails to ensure that experiments like this (which are real-life experiments, not far-fetched possibilities) produce determinate outcomes.

One might take cases like this to provide a reason to reject the GRW approach quantum mechanics, with or without the fuzzy link. But there are two reasons why such arguments aren't conclusive. First, when you observe a speck of dust or a flash on a screen, you correlate its state with the state of a much bigger object—your brain. Given the large numbers of particles involved in visual processing, the GRW dynamics can arguably ensure that your brain rapidly ends up close to an eigenstate of seeing the speck or flash in a determinate location, even if it didn't have a determinate location before you observed it (Aicardi, Borsellino, Ghirardi, & Grassi, 1991).[13] Second, even if your brain state doesn't immediately end up close to an eigenstate of some determinate visual experience, the arguments of the previous section entail that you won't notice; you will think that your experience is determinate. As long as such indeterminacy is not too frequent or long-lasting, it does not result in the semantic incoherence that was the downfall of the bare theory. So the problem here is not a straightforward empirical one. Still, it does mean that certain experiments will not immediately have a determinate outcome, and this is somewhat strange. In most cases, measurements acquire determinate outcomes due to physical processes in the external world, but in some cases the measurement only acquires a determinate outcome due to processes in the brain of the observer.

Along similar lines, the fuzzy link entails some peculiar properties for collections of macroscopic objects. Consider an object like a marble, and a location like being inside a certain box. According to the GRW theory, the state of the marble will generally be close to an eigenstate of location. We can represent such a state as follows:

$$a|\text{in}\rangle_1 + b|\text{out}\rangle_1, \tag{4.6}$$

where $|\text{in}\rangle_1$ is the eigenstate in which this marble (call it "marble 1") is located inside the box, $|\text{out}\rangle_1$ is the eigenstate in which it is located outside the box, and $|a|^2 + |b|^2 = 1$. As long as $|a|^2 > 1 - P$, the marble counts as being inside the box according to the fuzzy link. Now consider a large number N of marbles in similar states. The state of all the marbles together is

$$(a|\text{in}\rangle_1 + b|\text{out}\rangle_1)(a|\text{in}\rangle_2 + b|\text{out}\rangle_2)\ldots(a|\text{in}\rangle_N + b|\text{out}\rangle_N) \tag{4.7}$$

Looking at this state, we see that each marble individually counts as being in the box according to the fuzzy link, since each is sufficiently close to the relevant eigenstate. But now consider the N-marble state $|\text{in}\rangle_1|\text{in}\rangle_2\cdots|\text{in}\rangle_N$, which is the eigenstate for all the marbles being in the box; multiplying out the parentheses in (4.7) tells us that the coefficient on this term is a^N. Hence, the fuzzy link says that all N marbles are in the box if and only if $|a|^{2N} > 1 - P$. But since $|a|^2 < 1$, this condition is bound to be violated for a large-enough value of N. That is, for a large-enough collection of marbles, each marble individually is determinately located in the box according to the fuzzy link, but the collection of all N marbles does not have a determinate location (Lewis, 1997).

This problem has become known as the *counting anomaly* (Clifton & Monton, 1999, 700). How serious is it? The attitude of most commentators seems to be that if the counting anomaly were in fact manifested by sets of objects, it would be a serious problem for quantum mechanics with the fuzzy link, but in fact it is not manifested, so there is no problem. Clifton and Monton (1999, 711) take this position. They argue that as soon as someone observes the anomalous state, the anomaly disappears, since the GRW theory entails that if you correlate state (4.7) with the state of some macroscopic object (a counter or a brain), it rapidly evolves to a state close to an eigenstate of some determinate number of marbles being in the box. They conclude that there is no counting anomaly; an unobservable anomaly is no anomaly at all. But this is a bit quick. Because $|a|^{2N}$ is significantly less than 1, there is a good chance that the determinate number of marbles you find in the box on observation is less than N. Hence, it seems that observation doesn't make the anomaly go away; rather, it makes the anomaly manifest, since the premeasurement state (4.7) is one in which each marble is in the box according to the fuzzy link, and yet one is not certain (or even almost certain) to find them all in the box when you look (Lewis, 2003, 138).

Frigg (2003) takes a similar position, arguing that there is no counting anomaly because all it takes for our claim "All N marbles are in the box" to be true is just that each marble taken individually counts as being in the box. Since (4.7) is a state in which each marble individually counts as being in the box according to the fuzzy link, then it is a state in which all N marbles are in the box, notwithstanding its distance from the N-marble eigenstate $|in\rangle_1 |in\rangle_2 \cdots |in\rangle_N$. Furthermore, state (4.7) *behaves* as if all the marbles are in the box when we count them, since "counting is a process that is concerned with individual objects rather than with ensembles as a whole" (Frigg, 2003, 53).

However, it is not at all clear that Frigg is entitled to make these claims about ordinary language. Because we do not make the distinction in ordinary language between the claim that the ensemble of marbles is in the box and the claim that each of the marbles individually is in the box, it is implausible to argue that "All the marbles are in the box" means one rather than the other (Lewis, 2003, 140). Indeed, it is precisely this lack of a distinction that makes the counting anomaly seem paradoxical. On the one hand, state (4.7) is one in which each marble individually is in the box, and on the otherhand, it is one in which the ensemble has no determinate location, yet these two locutions are ordinarily taken to say the same thing. Similarly, the *behavior* of the marbles is paradoxical, because each individual marble behaves (with very high probability) as if it is in the box, and the ensemble does not behave (with high probability) as if N marbles are in the box, yet we do not ordinarily distinguish these cases.[14]

This latter formulation of the counting anomaly is reminiscent of epistemic paradoxes like the preface paradox (Makinson, 1965). Suppose that I believe a proposition if and only if my subjective probability in it is greater than 0.95. If I have 100 propositional beliefs each with a subjective probability of 0.98, then each proposition taken individually counts as a belief, but the conjunction of all 100 propositions does not. The preface paradox cannot be resolved by claiming that "I believe all the assertions" is about each individual assertion but not about their conjunction, since we do not ordinarily make such a distinction: "I believe all the assertions" is about each belief individually, but equally it is about the conjunction. Similarly, "All the marbles are in the box" is about the state of each marble individually, but equally it is about the joint state of the ensemble.

So I don't think one can argue that there is no counting anomaly; the counting anomaly follows from the structure of state (4.7) and the fuzzy link. If you want to avoid the counting anomaly, you have to make a more decisive break with the strict link: You have to reject it completely rather than liberalizing it. Bassi and Ghirardi (1999) take this route, invoking a modification to the GRW theory outlined by Ghirardi, Grassi, and Benatti (1995) according to which the quantum state does not constitute a complete description of the physical world. In addition to the quantum state, they postulate a continuous distribution of *mass* over space; the quantum state acts on this distribution, pushing it around so that it is dense in some places and rarefied in others. Roughly speaking, the mass density mirrors the probability density given by the square of the wave function amplitude: The mass density is high where the probability density is high. It is the mass distribution that directly describes the physical objects we interact with; marbles and coffee mugs are regions of relatively high mass density. This theory—GRW plus the mass density distribution—is sometimes called *massy GRW*.

Massy GRW has no need for the fuzzy link, since it eschews any direct link between the quantum state and physical properties. The condition for a marble being in a box makes no mention of the quantum state: The marble is in the box if and only if the associated region of high mass density is in the appropriate region of space. Even though a marble in a state like (4.6) has a tiny component of probability density outside the box, there is no associated region of relatively high mass density outside the box, since the tiny marble mass density outside the box is negligible relative to the mass density of the surrounding air. Because the counting anomaly is a product of the fuzzy link, this theory has no counting anomaly either: All N marbles are in the box if and only if there are N regions of high mass density in the box, that is, if and only if each marble individually is in the box.

So massy GRW successful avoids the counting anomaly. But at what cost? By distancing the quantum state from the physical description of the world, it also distances quantum behavior from the physical description of the world in a way that might be regarded as problematic. Bassi and Ghirardi (1999) count state (4.7) as a state in which all N marbles are in the box, despite the fact that if you were to count the marbles, the chance that you would find them all in the box is low. The trade-off for a solution to the counting anomaly is anomalous behavior. Furthermore, massy GRW is somewhat baroque. The fuzzy link was introduced to complement a particular modification to the Schrödinger dynamics: The modified dynamics ensures that the states of macroscopic objects are close to location eigenstates, and the fuzzy link ascribes determinate properties to objects in such states. This twofold strategy allows the quantum state to directly describe the systems we observe without the catastrophic indeterminacy of the bare theory. Massy GRW accepts the modification to the Schrödinger dynamics, but it denies that the quantum state directly describes the systems we observe. Hence, there is no need for the fuzzy link. But neither is there any great need for the modification to the Schrödinger dynamics. If the state itself doesn't describe the systems we observe, then solutions to the indeterminacy problem are available that do not require us to modify the Schrödinger dynamics—notably hidden variable theories like Bohm's. The postulation of a mass density in addition to the quantum state essentially makes massy GRW a hybrid collapse/hidden variable view, which seems unnecessarily complicated.

Fortunately, though, I don't think the counting anomaly *forces* us to take this route, because unlike the aforementioned authors, I don't take the counting anomaly to be a serious problem. Rather, it just indicates another failure in our ordinary language; it is not equipped to deal with states like (4.7). As noted earlier, for any two properties for which we have distinct terms (like "in" and "out"), there is a continuum of quantum states between the corresponding eigenstates. This leads to a continuum of behavior; the eigenstates always behave like systems with the relevant property, but the intermediate states sometimes behave like systems with one of the properties and sometimes like systems with the other. The binary distinction in our language—"in" and "out"—is not fine-grained enough to cope. We do have an adequate language for describing the system—it is the language of quantum states—but it doesn't map cleanly onto our ordinary language. Perhaps the fuzzy link is the best translation scheme we can come up with between the two languages; it gets as close as possible, given that the language of quantum mechanics can say things that can't be said cleanly in English.

But perhaps not; suppose we reconsider this whole business of marbles and boxes on the basis of the vague link rather than the fuzzy link. The vague link, recall, says that a system has a property to the extent that its state is close to the relevant eigenstate. So a marble in state (4.6) almost entirely has the property of being in the box but also has the property of being outside the box to a tiny extent. For all practical purposes, such a marble is in the box. Nevertheless, if you accumulate enough such marbles, even though each of them individually almost entirely has the property of being in the box, the ensemble of marbles has the property of being in the box to a much lesser extent. The ensemble cannot be said to be in the box for all practical purposes. The counting anomaly is still there, but now that property possession is conceived of as a matter of degree, it seems inevitable and benign.

The trick here, of course, is that the vague link, and its associated conception of property possession, is much closer to the quantum mechanical representation of the world, in which there is a continuum of states between the "in the box" state and the "outside the box" state. By the same token, it is much further from our ordinary binary property dichotomies, like "in" and "out." Taken as a translation scheme between quantum language and ordinary language, it is a failure, since ordinary predicates like "in" and "out" are not a matter of degree. But taken as a proposal about how to talk about the world in light of quantum mechanics, it is not bad—and arguably better than the fuzzy link. But either way, the counting anomaly does not seem fatal to the GRW project.

Where does this leave us regarding indeterminacy in the world? For microscopic systems, GRW-type theories all agree; microscopic systems can exhibit indeterminacy with respect to their properties, and this indeterminacy has nothing to do with compositionality or with familiar kinds of vagueness. Even small visible systems, like those underlying flashes on fluorescent screens, can exhibit indeterminacy of this kind—although as we have seen, this indeterminacy will not be noticeable by us. For macroscopic systems, there is of course ordinary compositional indeterminacy. But in addition, quantum indeterminacy can in principle reemerge at the macroscopic level: While individual macroscopic systems will (for all practical purposes) exhibit no such indeterminacy, nevertheless the indeterminacy would reemerge for (impracticably) large collections of macroscopic objects. And even though the macroscopic systems are composed of smaller components, the indeterminacy that emerges here has nothing to do with this compositionality; it has the same source as that affecting microscopic systems, namely that there is a continuum of quantum states between the eigenstates representing the determinate properties of the system, a continuum that requires no decomposition of the system into parts. But while this quantum indeterminacy has nothing to do with compositionality, it does involve vagueness—a new kind of vagueness, not vagueness about which microsystems are part of the macrosystem, but instead vagueness about where along the quantum continuum the system can be said to possess the relevant property. The vague link leaves this new vagueness exposed; the fuzzy link covers it with a sharp cutoff in the value of P, but the vagueness is still there, since there is presumably no sharp fact of the matter concerning the value of P.

4.4 Indeterminacy and Branching

Quantum indeterminacy has been most thoroughly investigated for GRW-type theories; the fact that GRW collapses put the quantum state *close* to a position eigenstate invites such considerations. But what of the other versions of quantum mechanics introduced in Chapter 3?

The consequences in the case of the many-worlds theory are very similar to those for the GRW theory. At the microscopic level, things are almost exactly the same: GRW collapses are sufficiently rare for isolated microscopic systems that for all practical purposes such systems behave as if there were no such collapses—that is, as if the many-worlds theory were true. So in the many-worlds theory, too, microscopic systems can have indeterminate properties, where this indeterminacy is primitive and has nothing to do with compositionality or familiar kinds of vagueness.

For macroscopic systems, the story is rather different in the two cases, because the behavior of macroscopic systems differs between them. In the GRW case, the collapse mechanism keeps the states of macroscopic objects close to position eigenstates, whereas in the many-worlds case there are no collapses and the states of macroscopic objects are generally far from position eigenstates. However, note that in the many-worlds theory (unlike the bare theory) it is the state of an object relative to a *branch* that determines its properties. So to figure out the extent of indeterminacy in the many-worlds theory, we first need to say a little more about branch structure.

Recall from Chapter 3 that branches are not fundamental entities according to the many-worlds theory; rather, the branching structure emerges as a result of the operation of the standard Schrödinger dynamics for sufficiently complex systems. The physical phenomenon that underpins the branching process is called *decoherence*. Consider first a single electron that passes through a two-slit interference device (Chapter 1). The two components of the wave function interact with each other after the slits to produce the characteristic probability distribution. When two wave components can come together to produce interference effects, they are described as *coherent*. Coherence, however, is fragile. Suppose that a second particle is close to the right-hand slit, so that the electron wave component that passes through that slit interacts with the wave of the particle. The result is that the wave component passing through the right-hand slit is shifted in the coordinates of the second particle, whereas the wave component passing through the left-hand slit is not. (Recall that the quantum state for a two-particle system occupies a six-dimensional space, in this instance consisting of three coordinates for the electron and three for the second particle.) If the shift is large enough, the two wave components now completely pass by each other in the region after the slits: Even though they overlap in the coordinates of the electron, they do not overlap in the coordinates of the second particle. This effectively destroys the interference; the second particle is said to *decohere* the two wave components.

Decoherence is a matter of degree in two distinct senses. First, if the interaction with the particle only shifts the right-hand wave component slightly, there may still be enough overlap between the two components for some interference effects to be observed. The smaller the potential interference effects, the more complete the decoherence. Second, it is relatively straightforward to reverse the decoherence in this case by undoing the effects of the interaction with the second particle on the right-hand wave component. But reversing decoherence is only feasible when the interactions of the electron with its environment are completely known. So the more complex and uncontrolled the interactions with the environment, the less likely it is that the decoherence can be reversed.[15]

The lesson is that if you want to observe interference effects, you need to keep the system in question isolated from outside influences, since even an interaction with a single particle can destroy the effect. This is difficult for microscopic systems, but effectively impossible for macroscopic ones. Such systems will inevitably interact with photons, air molecules, and other aspects of their environment in complicated and untraceable ways. Decoherence for macroscopic systems is rapid, very complete, and highly irreversible. This means that if the state of a macroscopic system comes to have two components, those components will not exhibit any appreciable interference effects. And since "interference" is just a name for interaction between two wave components, this means that for all practical purposes the two components do not interact. They can be regarded as separate branches of physical reality.

Even though the many-worlds theory has no collapse process, decoherence results in *effective* collapse—the appearance of collapse from within a branch. The Schrödinger dynamics typically results in the wave state of an object spreading out over time. But when this spreading out results in different parts of the wave interacting differently with the environment, then decoherence means that these components form separate branches. From the perspective of an observer within a branch, it looks as if the state of the object alternately spreads and narrows as the branching process takes place, much as in the GRW theory the state of a macroscopic object alternately spreads and narrows due to the collapse process. So branching mimics GRW collapse (for the inhabitants of the branches), and this means that the states of macroscopic objects are typically close to, but not exactly, eigenstates of position.

So something like the same problem arises within each branch under the many-worlds theory as arises globally under GRW. To see the nature of the problem, we first have to define the branching structure, and the simplest way to do that is to pick some property as the basis for the branching. For example, if a particular object has a different location in each branch, we can write the state of the system as

$$|\psi_1\rangle_s |l_1\rangle_o + |\psi_2\rangle_s |l_2\rangle_o + \ldots + |\psi_n\rangle_s |l_n\rangle_o, \qquad (4.8)$$

where the $|l_i\rangle_o$ are location eigenstates for the object in question and the $|\psi_i\rangle_s$ are the states of the rest of the system relative to the *i*th branch.[16] Then we can write a branch-relative version of the strict link as follows:

> **Branch-relative strict link:** A system has a determinate value for a given determinable property relative to a branch if and only if its state relative to that branch is an eigenstate of the operator corresponding to the property, and the determinate value is the eigenvalue for that eigenstate.

The problem is that the states $|\psi_i\rangle_s$ of the rest of the system will generally not be eigenstates of anything relevant to our experience, so according to the branch-relative strict link no object (apart from the one used to define the branches) has a determinate location.[17] And neither will the states $|\psi_i\rangle_s$ generally be states in which the observers in the branch are having determinate experiences. At first glance, then, the many-worlds theory plus the branch-relative strict link leads to exactly the same radical indeterminacy that was the downfall of the bare theory.

But in fact things are not quite so bad for many-worlds as for the bare theory. If we are free to choose the basis states $|l_i\rangle_o$ however we like, then each observer can choose them to be whatever states make her own experience determinate; presumably they will be complicated states of her own brain. This might look suspiciously ad hoc, but it can be defended: There is no objectively preferred branching structure, so as long as the pattern in which a particular observer has determinate experiences exists in the branching structure construed in some way or other, then we can say that the many-worlds branching structure contains those determinate experiences. In this way we can recover the determinate experience of each observer individually, each in their own slightly different construal of the branching structure.

Perhaps this is tenable, but it is far from satisfactory. The result is a kind of branch-relative solipsism: In the branching structure defined by your own determinate experiences, objects don't have determinate locations, and other people don't have determinate experiences. A better solution, as in the GRW case, seems to be to liberalize the strict link and adopt a branch-relative version of the fuzzy link or the vague link:

Branch-relative fuzzy link: A system has a determinate value for a given determinable property relative to a branch if and only if the squared projection of its state relative to that branch onto an eigenstate of the corresponding operator is greater than $1 - P$, where the determinate value is the eigenvalue for that eigenstate.

Branch-relative vague link: A system has a determinate value for a given determinable property relative to a branch to the extent that the squared projection of its state relative to that branch onto an eigenstate of the corresponding operator is close to 1, where the determinate value is the eigenvalue for that eigenstate.

Either of these links can ensure that objects have positions and people have beliefs relative to any reasonable specification of branching structure. The choice between the two links rests on essentially the same considerations as were in play for the GRW theory, namely the extent to which we are concerned that an object that has one determinate property might behave as if it has a different determinate property.

If we adopt one of these links, then, indeterminacy in the many-worlds theory has almost exactly the same extent as in the GRW theory. Differences between the two theories can occur, though, due to the reliance of the many-worlds theory on decoherence to define outcomes. Consider, for example, an electron whose position is measured by running it into a fluorescent screen. We saw that according to the GRW theory, if the electron lacks a determinate position (because it is in a superposition of distinct positions states), then the flash, too, will lack a determinate location on the screen. But the same does not necessarily follow for the many-worlds theory. The interaction with the screen can be regarded as decohering the terms in the electron's state, since each term becomes correlated with a distinct emission of photons from the screen. This argument is perhaps not decisive: Because only a few particles are involved in the interaction, the decoherence is not as irreversible as it could be. But certainly a case can be made that the interaction with the screen induces many-worlds branching, and hence that the flash on the screen occurs at a determinate location in each branch.

The other consequences of adopting the fuzzy link or the vague link noted earlier for the GRW theory apply in the same way to the many-worlds theory. Just as in the GRW theory, (impracticably) large systems of objects will be subject to the counting anomaly (relative to a branch), and while neither the fuzzy link nor the vague link makes the anomaly go away, each provides a clear way to talk about the states in question. So even though the underlying physical picture is different, the results concerning indeterminacy are surprisingly similar for the many-worlds theory and for GRW. Unanalyzable indeterminacy, having nothing to do with composition or standard vagueness, is endemic at the microscopic level, but for all practical purposes is absent at the macroscopic level—although it would reemerge for very large collections of macroscopic objects.

Now let us turn to Bohm's theory. As one might expect, matters are somewhat different here, since the theory postulates particles with precise determinate positions at all times, whatever the wave function state. So while macroscopic objects may be subject to compositional indeterminacy, they will not be subject to unanalyzable quantum indeterminacy, at least as regards their positions.

Bohm's choice of position as the property to make determinate is not accidental; arguably, at least, all our direct observations are observations of the position of something. In the paradigm case, the outcome of a measurement is given by the position of a pointer on a dial, but even in tricky cases like the measurement of the position of an electron using a fluorescent screen, the position of the electron and the positions of the photons emitted by the screen at the collision point are always determinate. So unlike the GRW theory, Bohm's theory entails that this measurement has a determinate outcome at the screen, prior to the involvement of the brain of the observer.[18]

But although the Bohmian strategy arguably makes all the properties we directly observe determinate, it does not thereby make *all* properties determinate. Consider, for example, an electron located in spatial region o, in a superposition of z-spin eigenstates:

$$\frac{1}{\sqrt{2}}\left(|o,\uparrow\rangle_e + |o,\downarrow\rangle_e\right). \tag{4.9}$$

The electron has a determinate position but not a determinate spin, since the position of the Bohmian particle (somewhere in o) does not pick out one spin property or the other. If we correlate its position with its spin, its final state will be

$$\frac{1}{\sqrt{2}}\left(|+,\uparrow\rangle_e + |-,\downarrow\rangle_e\right), \tag{4.10}$$

where $+$ and $-$ are two distinct regions of space. Now the position of the Bohmian particle, either in region $+$ or region $-$, does determine the spin of the electron. So since spin measurements involve correlating the spin of the electron with its position, the result of the measurement reflects a determinate spin property that the electron has after measurement, but not a determinate spin property that the electron had prior to measurement.

But even this description involves an important idealization. Because no real process can *perfectly* correlate the electron's spin with its position, the postcorrelation state will actually be more like this:

$$\frac{1}{\sqrt{2}}\left(a|+,\uparrow\rangle_e + b|+,\downarrow\rangle_e + a|-,\downarrow\rangle_e + b|-,\uparrow\rangle_e\right), \tag{4.11}$$

where $a^2 \gg b^2$. That is, if the particle is in region $+$, the state of the electron will be close to, but not exactly, an eigenstate of z-spin, and similarly for region $-$. So if we want to ascribe a determinate spin to the electron after the spin measurement, we have to appeal to something like the branch-relative fuzzy link or the branch-relative vague link defined earlier.

So the situation for properties other than position in Bohm's theory is very like the situation in the many-worlds theory. But there are some important differences. First, the choice of basis for writing down the branching structure is made for us in Bohm's theory: It has to be position, since it is the positions of the Bohmian particles that determine all the other properties. Second, there is no requirement that the branches be decoherent in the Bohmian case. Suppose the two terms in state (4.10) later come together and interfere; still, the Bohmian particle follows one branch or the other before the interference occurs, and hence the electron has a determinate spin value at that time, even if the interference means that later on we cannot find out what that determinate value was.

So Bohm's theory doesn't purge the world of unanalyzable quantum indeterminacy at the microscopic level, but combined with the fuzzy link or the vague link, it makes the indeterminacy go away when we look for it. Perhaps more surprisingly, it doesn't make the indeterminacy at the macroscopic level go away entirely either. The counting anomaly as formulated earlier doesn't affect Bohm's theory because that formulation is in terms of the positions of the marbles, and no appeal to the fuzzy link is required for position in Bohm's theory. But the counting anomaly can be reformulated in terms of some other property. Suppose instead that each marble is in a superposition of distinct energy eigenstates, $a\,|\text{low}\rangle_i + b\,|\text{high}\rangle_i$, where $a^2 \gg b^2$. The location of the Bohmian particles doesn't pick out one of these terms as actual, since (we will suppose) the energy of the marble is not correlated with its position.[19] But we can apply the (branch-relative) fuzzy link to each particle individually, and it tells us that each marble has a determinate energy, namely the low one. However, since the coefficient on the term $|\text{low}\rangle_1\,|\text{low}\rangle_2\,...\,|\text{low}\rangle_N$ is a^N, which will be small for sufficiently large N, the fuzzy link applied to the collection of marbles tells us that it is not the case that all the marbles have determinate energy. The anomaly is not as striking as for the GRW theory, as energy is not a directly observable quantity, but it is still there behind the scenes.

So there are some differences concerning the scope of indeterminacy in the world according to GRW, many-worlds, and Bohm. But it is notable what they have in common: Widespread indeterminacy at the microscopic level that disappears at the macroscopic level. Perhaps here we have our first concrete ontological consequence of quantum mechanics: There is indeterminacy in the world, and it has nothing to do with composition or (familiar kinds of) vagueness.

4.5 Avoiding Indeterminacy

But perhaps not. In fact, there are two distinct ways one might try to avoid positing indeterminacy in the quantum world. One possible strategy is to extend the Bohmian hidden variable approach to properties other than position. The immediate problem with this strategy is that the no-go theorems explained in Chapter 2 apparently rule it out: The Kochen-Specker theorem tells us that any attempt to assign determinate values to *every* determinable property lands us in contradiction. But we noted in that chapter that the theorem relies on assumptions that, though extremely plausible, might be violated. In particular, by violating the assumption that the properties of the system are statistically independent of any measurements that will be performed on it later, one can avoid the conclusion of the Kochen-Specker theorem. The most plausible way to do this is to allow that later events can causally influence earlier events, reversing the usual direction of causation (Price, 1994).

If one takes this "retrocausal" approach, then there is no mathematical barrier to producing a hidden variable theory according to which every determinable property of a system always has a determinate value. Postulating backward causation might be seen as a fairly desperate means to this end. However, Price (1996) argues that it is only a temporal double standard that has prevented us from recognizing backward causation in the microworld, even aside from quantum mechanics. Nevertheless, there are obvious difficulties involved in positing backward causation, to be explored in Chapter 5. Furthermore, it is important to note that no explicit retrocausal hidden variable theory has been constructed, so until a fully developed theory along these lines becomes available, this is at best a promissory note.

The retrocausal strategy attempts to eliminate indeterminacy by postulating additional determinate property values not represented in the usual versions of quantum mechanics. But the opposite approach is also possible—we could limit the determinable properties we attribute to physical systems to those represented in our favorite quantum theory. This approach is most obvious for Bohm's theory. As noted earlier, position plays a special role in Bohm's theory: The positions of particles are always determinate, and consequently so are the positions of macroscopic objects, at least setting aside worries about compositional indeterminacy. So suppose we insist that positions are the only determinable properties that things really have. Particles don't really have spins, and spin-talk is really about how the positions of particles would behave in certain circumstances. If that strategy is successful, then there is no special quantum indeterminacy in the world—no indeterminacy over and above combinatorial indeterminacy, if one is prepared to admit that.

But as it stands this simple proposal isn't tenable. Recall that the particles are not the only element in the Bohmian model of microphysics. There is also a wave function that evolves over time and pushes the particles around. As noted in Chapter 3, it is hard to deny that the wave function represents a real entity.[20] So we at least need to include the properties of this entity. But while a spread-out wave function may not be interpretable as ascribing determinate values of familiar properties like energy or momentum to a system, the intrinsic properties of the wave function itself—its degree of spread and its shape—are perfectly determinate.

This is an attractive way to interpret Bohm's theory. The wave function has a perfectly determinate distribution, and the particles have perfectly determinate positions, and that's all the determinable properties there are. We can talk about a system's energy if we like, but this is just a useful shorthand for describing wave function properties and position properties in certain situations. The fact that the energy of a system is sometimes "indeterminate" just reflects the fact that energy talk is inapplicable for certain states, not that there is any real indeterminacy in the world.

In fact, this general way of approaching the properties ascribed to systems by quantum mechanics can be extended to many-worlds and GRW, too. In these latter cases, there are no determinate particle positions, so we just have the determinate properties of the wave function. Position talk, too, is just a useful way of talking about the wave function in these theories. When the wave function bunches up in three of its dimensions and remains bunched up, then it is useful to talk of a particle following a continuous trajectory through space. For more spread-out wave function states this kind of position talk fails, but not because objects have indeterminate positions. Rather, position is not a genuine determinable property of things.

If this is the right way to ascribe properties in the three main quantum theories, it suggests that we have been looking at things all wrong. We have been trying to interpret the quantum state in terms of familiar properties, and then we are forced to say that those familiar properties may have indeterminate values in the quantum realm. But instead, perhaps what we should have been doing is trying to interpret quantum states on their own terms—in terms of the properties of the wave function itself. After all, it is a commonplace that the quantum world is *unfamiliar*. Perhaps it was naïve of us to expect to find anything like the familiar properties of experience at the quantum level of description. After all, we don't expect determinables like *color* to apply to electrons; why expect *position* to apply to them either?

4.6 Conclusion

So does quantum mechanics introduce a new kind of indeterminacy in the world? While the debate is far from conclusive, it looks to me like it does, unless a retrocausal theory comes along to save the day. Even if we restrict determinable properties to direct properties of the wave function as suggested in the previous section, some of them will be indeterminate. Consider again the cloud-like electron of Figure 4.1. The picture is essentially a picture of wave function amplitude. The amplitude distribution that generates this picture is certainly determinate. But what about the *size* of the cloud? It seems hard to deny that the determinable *size* applies to the cloud; it's a spatially distributed entity. But due to the fuzzy nature of the boundaries—of the wave function tails spreading to infinity—there doesn't seem to be a determinate answer to the question of how big it is. This indeterminacy is not compositional indeterminacy, and hence it is of a new kind. However, the vagueness involved in this example *is* just the familiar kind of vagueness: The wave function has a continuum of amplitude properties, and the indeterminacy arises because we don't know where in the continuum to put the boundary of the cloud.

The existence of more widespread quantum indeterminacy—the kind that arises even for discrete properties—is harder to establish. There seems to be little reason to shun such indeterminacy: The fuzzy link and the vague link fit well with the established versions of quantum mechanics and don't lead to insurmountable problems. But as we have seen, it is not forced on us. Still, even if quantum mechanics doesn't resolve the debate, it opens our eyes to a number of unforeseen options concerning the nature of indeterminacy in the world, and it presents us with interesting empirical considerations that bear on the question of the existence of such indeterminacy.

5| Causation

ACCORDING TO BOHR, QUANTUM MECHANICS entails "the necessity of a final renunciation of the classical ideal of causality" (1935, 697). What he meant by this isn't very clear: The reason he gives for this conclusion is "the impossibility of controlling the reaction of the object on the measuring instrument," but this looks like a purely epistemic limitation.[1] Given the overall context of the paper though—a response to Einstein, Podolsky, and Rosen (1935) on the completeness of quantum mechanics—it seems likely that he has in mind something like the lack of a unique description of the microphysical realm. As we saw in Chapter 2, Bohr thinks that the properties of a quantum system depend on the measurement context in which it is placed. In that case, because we can choose which measurements to make at some later time, the system itself may have no well-defined properties, in which case no causal story about the provenance of the eventual measurement results can be given.

However, Bohr is certainly not the last word here. As we saw in Chapter 3, several different quantum theories can be constructed which *do* provide causal stories at the microphysical level. Still, Bohr may have a point, insofar as these causal accounts fall short of the "classical ideal" to which he refers in earlier. As we have seen, no-go theorems like Bell's place important restrictions on the kinds of properties that can be ascribed to systems, and hence on the kinds of causal stories we can tell. The causal stories given by hidden variable theories, collapse theories, and many-worlds theories have to respect these restrictions, and this may well mean that their accounts of measurement results deviate from an ideal of causal explanation you might extract from classical mechanics.

At first glance, though, the kinds of causal explanation given by the main interpretations of quantum mechanics look just like those given in classical physics. All three interpretations—Bohm, GRW, and many-worlds—explain measurement outcomes (at least in part) in terms of the evolution of the quantum state according to the Schrödinger equation. The Schrödinger equation has the form of a wave equation, so the explanation looks analogous to those given in classical electromagnetism—in terms of the propagation of waves in a field.[2] Explanations in Bohm's theory also add the motion of particles according to the Bohmian dynamical equation; but again, causal explanation in terms of the motion of particles is entirely familiar from classical mechanics.

So superficially, at least, the causal explanations given by the major interpretations of quantum mechanics seem to fit the "classical ideal" very well. But as we shall see, the causal explanations offered by each interpretation involve significant nonclassical causal features—features that are unfamiliar from the perspective of classical physics. The goal of this chapter is to uncover the causal structure of the microworld according to the main interpretations of quantum mechanics. Note that I make no attempt to show how (or whether) quantum mechanics is relevant to the *analysis* of causation. I presuppose that the explanations just sketched qualify as causal, and I assume that any reasonable analysis of causation would concur. So while it is possible that quantum mechanics has something interesting to tell us about causation as such, that is not the order of business here.

5.1 Locality

An important feature of many of the causal explanations given by classical physics is that they are *local*, in the sense that what happens to the system next depends only on its intrinsic properties and its immediate environment. For example, in classical particle mechanics, the motion of a particle depends only on its intrinsic properties (e.g., mass and charge) and its immediate environment (e.g., the electromagnetic field at the location of the particle, or immediate contact with other particles). For waves, the change in wave amplitude in a given location depends only on the properties of the field at that location.

A glaring exception to local causal explanation in classical mechanics is *gravity*, according to which the subsequent motion of a particle with mass depends on the locations of every other particle with mass right now, however distant they may be. But the nonlocality of classical gravitational explanation is typically regarded as a *problem*, and modern theories of gravitation attempt to make it a causally local phenomenon, too, so that gravitational influences propagate through space in much the same way as electromagnetic influences. The reason that causal nonlocality is regarded as problematic is that causal locality is arguably required by special relativity. So let us take a brief look at what special relativity tells us about the space-time structure of the world.

It is worth starting out by comparing prerelativistic views of space and time. Newton famously held that there is an objective distinction between motion and rest. But as Leibniz pointed out, any such distinction would be empirically idle (Huggett, 1999, 150). We can visualize the difference between these two views in terms of our freedom to choose coordinates for space and time, as shown (for one spatial dimension) in Figure 5.1. According to Newton, only the trajectory of an object that is genuinely at rest can serve to define the time axis t (the set of points where $x = 0$) in our coordinate system. But according to Leibniz, any inertial (nonaccelerating) trajectory can equally well serve as the time axis, so t and t' (the set of points where $x' = 0$) are equally good choices. According to Newton, trajectory AB depicts an object that is objectively at rest. But according to Leibniz, the object is at rest under one choice of coordinates (A and B have the same x-coordinate) and in motion under the other choice (A and B have different x'-coordinates), so its state of rest or motion is entirely conventional.

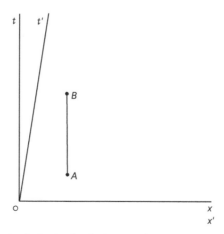

FIGURE 5.1 Coordinate choice in classical space-time.

This all seemed fine until we ran into the recalcitrant fact that the speed of light is the same relative to every observer. This sets light apart: A car that is traveling at 70 km/h relative to the road, for example, is traveling at 140 km/h relative to a car traveling at 70 km/h in the opposite direction, but a light-beam traveling at c relative to the road is still traveling at exactly c relative to a car traveling in the opposite direction, no matter how fast the car is going. To accommodate the anomalous behavior of light, Einstein suggested that a change in our choice of time axis needs to be accompanied by a corresponding change in our space axis; this is his special theory of relativity. The situation is shown (again for one spatial dimension) in Figure 5.2. If we choose time axis t' instead of time axis t, then we also need to switch our space axis from x (the set of points where $t=0$) to x' (the set of points where $t'=0$). Note the symmetry here: While in general the speed of an object depends on the choice of coordinates, for an object traveling at the speed of light—the dashed diagonal trajectory—the speed is the same in the primed and unprimed coordinates. The light travels less far in the primed coordinates, but it also takes less time, and these factors exactly cancel out. Now consider events A and B in Figure 5.2. Under one choice of coordinates, they happen at the same time (they have the same t coordinate), but under another choice B occurs before A (they have different t' coordinates). So just as rest is entirely conventional according to (Leibnizian) classical mechanics, so simultaneity is entirely conventional according to special relativity.

Now we can see precisely why causal locality is required by special relativity. Suppose the motion of a particle depends on the value of some

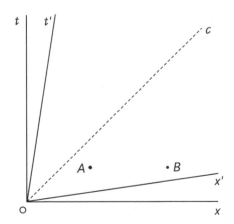

FIGURE 5.2 Coordinate choice in special relativity.

distant property right now—say, the mass of a distant star. According to special relativity, whether two events are simultaneous is a matter of conventional choice, not objective fact. So the value of a distant property *right now* is simply undefined according to special relativity. The motion of a particle here and now cannot depend on distant states of affairs, not because such instantaneous action at a distance would be "spooky," but because there is no objective fact about which distant states are simultaneous with here and now. Only local causal explanations are allowed, because only local causal explanations are well defined.

How does quantum mechanics fare when it comes to special relativity? Recall that one of the assumptions of Bell's theorem (Chapter 2) is also called "locality." How is this assumption related to the causal locality we have been considering here? Bell's locality assumption says that for a pair of spatially separated particles, a measurement on one particle cannot instantly affect the properties of the other. Clearly any process that allows instantaneous causation at a distance is causally nonlocal: The behavior of the affected particle does not depend only on its intrinsic properties (prior to the measurement) and its immediate environment. So if we want to get around Bell's theorem by violating his locality assumption, we need to posit a causally nonlocal theory.

So at least some ways of getting around Bell's theorem—those that violate the locality assumption—are in conflict with special relativity. If the theory in question says that the subsequent behavior of an object depends on what is going on right now at some distant location, then the theory is prima facie incoherent, because "right now" cannot be objectively defined for spatially separated points. This means that some causal stories that appear to be good explanations may turn out to be unacceptable after all. To what extent the various interpretations of quantum mechanics suffer from this problem, and to what extent it can be remedied, are taken up in the following sections.

5.2 Particle Trajectories

Let us begin by considering the case of Bohm's theory, because the causal nonlocality is particularly striking here.[3] Consider the state involved in Bell's theorem—the entangled spin state $|S\rangle = (|\uparrow\rangle_1 |\downarrow\rangle_2 - |\downarrow\rangle_1 |\uparrow\rangle_2)/\sqrt{2}$ for two spin-1/2 particles, where the spins are relative to the z-axis. The particles (and their associated wave packets) are spatially separated (say along the x-axis), and each particle in turn has its spin measured in the z direction.

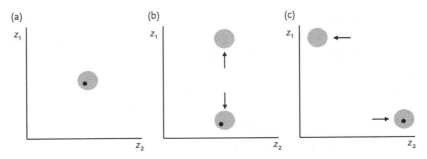

FIGURE 5.3 Entangled spins: Particle 1 measured first.

The way in which Bohm's theory describes this process can be illustrated as in Figure 5.3. The vertical axis represents the z-coordinate of particle 1, and the horizontal axis represents the z-coordinate of particle 2. The shaded areas represent the regions in which the wave function amplitude is large, and the dot represents the positions of the two Bohmian particles (the vertical coordinate representing the position of particle 1 and the horizontal coordinate the position of particle 2). The initial state of the two-particle system is as shown in Figure 5.3 (a): The wave function in the coordinates of each particle is large only within the same small range of z values (although the wave packets are widely separated along the x-axis), and each of the particles is somewhere in this region.

Suppose we measure the spin of particle 1 by passing it through a magnetic field oriented along the z-axis. This has the effect of dividing each wave packet vertically into two according to the spin of particle 1: The $|\uparrow\rangle_1|\downarrow\rangle_2$ component of the wave packet moves up, and the $|\downarrow\rangle_1|\uparrow\rangle_2$ component moves down, as shown in Figure 5.3 (b). Particle 1 is carried with one or the other packet, according to the Bohmian dynamical law. An important fact about this law is that it is deterministic, and this means that the various possible trajectories for the particle cannot intersect, because otherwise the behavior of the particle would be indeterministic at the intersection point. In this case, it means that if the particle is in the upper half of the wave packet it is carried upward, and if it is in the lower half it is carried downward—because otherwise the possible trajectories would cross each other. In Figure 5.3 (a), the particle is initially in the lower half of the packet, so it moves down, and hence is measured as spin-down. If we subsequently measure the spin of particle 2, then the $|\uparrow\rangle_1|\downarrow\rangle_2$ component of the wave packet moves down in the coordinates of particle 2, and the $|\downarrow\rangle_1|\uparrow\rangle_2$ component moves up, as shown in Figure 5.3 (c). Due to the perfect

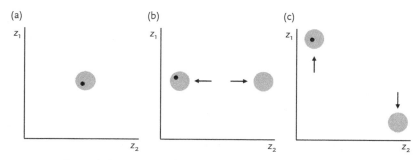

FIGURE 5.4 Entangled spins: Particle 2 measured first.

anti-correlation of the spins in this state, there is no further splitting of the wave packets. So in this case, since particle 1 is carried downward, particle 2 is carried upward, and is measured as spin-up.

But suppose instead that the spin of particle 2 is measured first. Then each wave packet initially divides into two components in the z-coordinate of particle 2, as shown in Figure 5.4: The $|\uparrow\rangle_1|\downarrow\rangle_2$ component moves down, and the $|\downarrow\rangle_1|\uparrow\rangle_2$ component moves up. In our case, because particle 2 is initially in the lower half of the wave packet (relative to the z_2 coordinate), it moves down and is measured as spin-down. If we subsequently measure the spin of particle 1, then the $|\uparrow\rangle_1|\downarrow\rangle_2$ component of the wave packet moves up in the coordinates of particle 1, and the $|\downarrow\rangle_1|\uparrow\rangle_2$ component moves down. So in this case, since particle 2 is carried downward, particle 1 is carried upward, and is measured as spin-up.

This is an odd situation. Particle 1 is measured as spin-down if its spin is measured first, but as spin-up if particle 2 has its spin measured first. One might well wonder what the premeasurement spin of particle 1 *actually* is, but it is worth recalling that Bohm's theory doesn't directly ascribe spin properties to particles, only positions. For the moment, the crucial point is that the outcome of the spin measurement on particle 1 depends on a distant event. If the two measurements are close enough in time and far enough apart in space, special relativity entails that there is no fact of the matter about which event occurs first: Under some choices of coordinates the measurement on particle 2 is performed before the measurement on particle 1; under other choices it is not. So since the dynamics of Bohm's theory depends on there being such a fact, the dynamics is simply ill defined according to special relativity.

So is Bohm's theory simply ruled out by special relativity? Here it is worth noting that the conflict is not a straightforward *empirical* defect in Bohm's theory. Recall that in Bohm's theory the position of a particle within the wave packet cannot in principle be known with more precision than given by the Born rule. So in this case, that means that there is a 50% probability that particle 1 will go up, regardless of whether particle 2 is measured first. So although whether or not a measurement has been performed on particle 2 affects the outcome for particle 1, it does not make any difference to what we should expect. Because we cannot *place* the particle at a particular location in the wave packet, we cannot empirically demonstrate the violation of locality shown in Figures 5.3 and 5.4.

Still, a conflict is a conflict, even if it has no empirical consequences. Special relativity is one of the pillars of 20th-century physics, and if it renders an interpretation of quantum mechanics incoherent, then one might reasonably think that the interpretation is untenable. But it is worth considering what it would take to make the Bohmian dynamics well defined. What we would need is an absolute standard of simultaneity—a standard that defines whether two spatially separated events are simultaneous or not—that we can appeal to in cases like the one just discussed to decide which measurement is performed first. In terms of Figure 5.2, we would need to pick one axis—$t = 0$ to $t' = 0$ or some other—as privileged.

But note that the specification of an absolute standard of simultaneity does not affect any empirical predictions. Just as in classical mechanics Newton's concept of absolute rest has no empirical consequences, so in relativistic mechanics a putative absolute standard of simultaneity would have no empirical consequences. Einstein follows Leibniz's methodological conviction that if an element of a physical theory would make no empirical difference, we should leave it out. However, it is not as if special relativity *rules out* the existence of an absolute standard of simultaneity; rather, it declares such a standard otiose. From that perspective, things don't look so bad for Bohm's theory. Even if an absolute standard of simultaneity has no direct empirical use, it might find an indirect use in allowing a hidden-variable understanding of quantum mechanics. After all, if we are in the business of modifying standard quantum mechanics (which Bohm's theory certainly does), then there is no clear reason not to modify special relativity while we are at it. This strategy has been explored by Bell (1976a) and defended by Maudlin (1994).

If you are willing to bite the bullet here, our standard picture of causation has to be modified by the addition of nonlocal causal influences—"spooky action at a distance," in Einstein's memorable phrase (Born, 1971, 155). Despite Einstein's qualms, this is a fairly modest conceptual change, and it is a mode of explanation with which we are already quite familiar from thinking about the influence of the planets on each other according to Newton's theory of gravitation.

So Bohm's theory may be tenable if you are willing to modify special relativity to match. However, this strategy may strike you as methodologically unwise: Many commentators find the modification of standard quantum mechanics required by Bohm's theory to be already ad hoc (e.g., Polkinghorne, 2002, 55), and the modification to special relativity arguably just compounds the problem.[4] So is it possible to construct a hidden variable theory that is consistent with special relativity? Well, locality is not the only assumption of Bell's theorem: Bell also assumes that the properties of a particle are independent of the measurements to be performed on it. As mentioned in Chapter 3, it may be possible to build a fully local hidden variable theory if one is willing to violate this independence assumption. Given the costs associated with violating locality, this is an interesting possibility to consider. So let us think about what the causal structure of such a theory would have to look like.

Bell himself considers violations of independence in order to escape the conclusion of his theorem, but reports that he is unable to take such a view seriously because it entails that "apparently separate parts of the world would be deeply and conspiratorially entangled, and our apparent free will would be entangled with them" (Bell, 2004, 154). Recall the set-up of Bell's theorem from Chapter 2. Two spin-1/2 particles in an entangled state can each have their spin measured in one of three directions 120° apart. If the spins are measured in the same direction, they never agree, but if they are measured in different directions, they agree 3/4 of the time. How can we ascribe spin properties in each of the three directions to each of the particles? If we respect locality and independence, we saw that the assignment of spin values is impossible. If we violate locality, we can complete the spin assignment by making the spin of particle 2 depend on whether a spin measurement is performed on particle 1; we just saw how this works in the case of Bohm's theory.

What if we violate independence instead, so that the spin values of each particle depend on which spin measurements will be performed on it? Given this dependence, it is straightforward to make the spin values more likely to agree if the two particles will be measured in different directions, generating the observed quantum mechanical statistics. But *how* could the spin of a particle depend on the measurements to be performed on it? Consider the space-time diagram of the Bell experiment in Figure 5.5 (a), where space is represented by the horizontal axis and time by the vertical axis. Here the diagonal lines are the moving particles, and the solid vertical

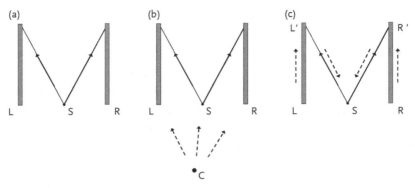

FIGURE 5.5 Space-time diagram of the Bell experiment.

bars are the measuring devices. S is the particle emission event, and L and R are the measurement choice events on the left and the right. Note that the measurement choice events are simultaneous with the particle production event (under this choice of coordinates). So the measurement choices can't directly affect the spins of the particles, or vice versa, without violating local causation.

One way a causal explanation of the correlation could be constructed is by postulating a common cause for the particle properties and the measurement choices, as shown in Figure 5.5 (b). Here there is an event C in the common past of S, L, and R that can affect them all. However, it is hard to see how this could work, since it seems fairly straightforward to *ensure* that the measurement chosen is independent of the particle properties, for example by choosing the measurement using a random number generator of some kind. Bell (2004, 154) suggests a Swiss national lottery machine or a free-willed agent.[5] For the common cause mechanism to be sufficient to overcome all such attempts to undermine the correlation, it must affect *anything* that might be used to choose the measurement. That is, there would have to be a vast hidden causal order, a massive physical conspiracy, enforcing the quantum correlations required to get around Bell's theorem (Lewis 2006a, 367).

So appealing to a past common cause is wildly implausible. But a second possibility is to postulate a cause in the *future* that explains the correlation. At first glance, this possibility seems even less plausible than the first, since it requires "backward causation" via which a future cause can produce a present effect. But a couple of things can be said in its favor. First, backward causation in microphysics can be independently motivated via the temporal symmetry of the underlying laws (Price, 1996). Second, the appeal to a cause in the future does not require a massive physical conspiracy: The particle necessarily interacts in the future with any device that performs measurements on it, so the act of measurement itself can act as the cause of the prior spin of the relevant particle. This is depicted in Figure 5.5 (c): The choice of measurement L causes the later measurement event L', and L' causally influences the *earlier* particle production event S (and similarly on the right).

This is the retrocausal hidden variable approach mentioned in passing in previous chapters. It is an intriguing possibility, but it needs to be treated with some care. As the arrows in Figure 5.5 (c) indicate, the postulated causal influences form *loops*—for example, from S to L' and from L' back to S—and causal loops are notoriously problematic. One kind of worry is that the loops trivialize the causal explanation of the measurement outcomes. Suppose that the way that the retrocausal mechanism is implemented is that the measurement setting at L' causes the particle at L' to have a determinate spin property—say spin-up in the z-direction—and this property of the particle has a (retro-)causal influence on the state of the particle at S. Then the measurement outcome at L' (retro-)causes the particle at S to be spin-up in the z-direction, and the spin of the particle at S causes the measurement outcome at L'. Our causal "explanation" goes in a very tight circle and is arguably no explanation at all (Lewis, 2006a, 376). What's more, this circle is self-contained: Since the spin of the particle at S is fully determined by the later measurement outcome, there is no role for the quantum state at S, and hence no obvious way to explain the statistical distribution of particle spins (Berkovitz, 2008, 718).

One can avoid these difficulties by supposing that only the measurement *direction* at L' is retrocausally effective. That is, the direction in which the measuring device is aligned at L'—say along the z-axis—is retrocausally transmitted to S, and the actual spin property of the particle—either spin-up or spin-down along the z-axis—is transmitted from S to L' (Lewis, 2006a, 376). Now the causal explanation is no longer trivial. But further worries can arise concerning the *consistency* of the causal loops. In particular, Berkovitz (2008, 712) envisions an experiment in which the measurement at L' occurs after the measurement at R', and the measurement outcome at R' is used to choose the measuring device setting at L'. The worry is that one could arrange things such that the spin properties of the particles at S produce a measurement outcome at R', and hence a measurement setting at L', which is inconsistent with the spin properties at S. For example, perhaps the spin properties at S are such that particle 1 is z-spin-up and particle 2 is z-spin-down, but a z-spin-down outcome at R results in L being set along the x-direction, and hence particle 1 having a determinate x-spin property but no determinate z-spin property.

An easy way out of *this* problem is to deny that the particle properties at S are simply randomly distributed according to the quantum state at S; rather, the range of allowed particle properties depends on future measurements. This is perfectly coherent in a retrocausal theory, but note that again it divorces the distribution of particle properties from the quantum state—the properties are not simply distributed at random according to the squared amplitude of the relevant terms in the state. We have no explanation of the agreement of the outcomes with the statistical predictions of quantum mechanics (Berkovitz, 2008, 717).

Of course, until we have an explicit retrocausal theory in front of us, we don't know whether these problems will arise in the form in which Berkovitz considers them: We don't know whether such a theory will retain either the quantum state or the Born rule in their standard forms. Several research programs currently aim at developing such a theory. Aharonov and Vaidman (1990) suggest a theory with two wave functions: The standard one evolving forward in time, plus a new one evolving backward in time. The status and nature of the backward-evolving state, and its relationship to the observed properties of systems, are matters of ongoing debate (Vaidman, 2010). Along similar lines, Sutherland (2008) constructs a retrocausal Bohmian model, in which the particle positions depend on both a forward-evolving and a backward-evolving wave function. Wharton (2010) suggests that we should represent quantum systems using a classical wave—that is, a wave residing in an ordinary three-dimensional space, rather than the quantum wave function in a high-dimensional space—where the classical wave is constrained by both past and future interactions. Price (2012) suggests a discrete ontology rather than a wave, where a quantum particle has two sets of properties, one constrained by its past interactions and the other constrained by its future interactions.

It should be clear that the retrocausal options are diverse; it is too soon yet to say whether the program will succeed, and if so, what the form of the final theory will be. If it succeeds, it rescues one aspect of our standard causal picture, namely causal locality, at the expense of another, namely the assumption that causes precede their effects. While the latter is arguably a more radical violation of our causal intuitions than the former, it has the advantage that it allows us to retain special relativity as it stands. But either way, the particle trajectories involved in hidden variable theories violate some aspect of the "classical ideal."

5.3 Wave Packets

Particle trajectories are not the only causal elements of Bohm's theory; the wave function itself evolves over time according to a causal law, as depicted by the circles in Figures 5.3 and 5.4.[6] And in theories without particle trajectories, like GRW and many-worlds, the wave function is the primary dynamical entity. As mentioned earlier, the causal process here is the propagation of a wave-like disturbance in a field. How closely does this causal process adhere to the classical ideal?

Let us start by considering causal locality. Consider again the spin measurements on entangled particles shown in Figures 5.3 and 5.4. Note that if we ignore the particles in these diagrams, the order of the measurements makes no difference to the final state; in each case there is a wave packet in the top left and a wave packet in the bottom right. The paths via which the wave packets get to this point *do* differ depending on which measurement is performed first, but we can regard this as an artifact of the way we have chosen our coordinates. That is, if we choose coordinates such that the measurement on particle 1 happens first, then it looks like the wave packet initially splits in the coordinates of particle 1, and if we choose coordinates such that the measurement on particle 2 happens first, then it looks like the wave packet initially splits in the coordinates of particle 2, but really there is no fact of the matter about which process happens first.

This suggests that there is no causal nonlocality in the wave portion of Bohm's theory—that the nonlocal causation is all in the particle trajectories. This is significant for the evaluation of the many-worlds theory, since the many-worlds theory is just Bohm's theory without the particles. If the causal nonlocality in Bohm's theory is all in the particles, then it looks like the many-worlds theory is causally local. But this suggestion needs to be treated with some care. The Schrödinger equation that governs the wave motion is local in the space in which the wave function lives, but recall that this space is not ordinary three-dimensional space, but a space that has three dimensions for each particle in the system. The diagrams in Figures 5.3 and 5.4 show only one spatial dimension (the z-axis) for each particle, and the circles really represent one region of space in the coordinates of particle 1 and a different region of space in the coordinates of particle 2 (since the wave packets are widely separated along the x-axis). The Schrödinger equation is local in the sense that the wave motion at a point in this six-dimensional space depends only on the properties of the wave function at this point—but the point in question in fact refers to two different locations in three-dimensional space.

So the account of wave motion generated by the Schrödinger equation is not causally local in ordinary three-dimensional space. Nor is it compatible with special relativity, for the same reason: The wave motion right now depends on the wave function properties of two distinct points in space right now, but "right now" has no objective meaning for distinct points of space according to special relativity. Perhaps this should not be a surprise, since the theory we have been working with is *nonrelativistic* quantum mechanics. Relativistic versions of the Schrödinger wave equation can be derived—the Dirac equation and the Klein-Gordon equation—which reproduce the Schrödinger equation in the nonrelativistic limit. The lack of causal locality inherent in the Schrödinger equation may just be an artifact of a particular nonrelativistic presentation.

So as far as we can tell, then, there is no essential causal nonlocality involved the many-worlds theory; we have no reason to think that the many-worlds theory exhibits any causal nonlocality that wouldn't go away in a rigorous relativistic treatment.[7] This gives the many-worlds theory a significant advantage over Bohm's theory: While we can accommodate nonlocal causation within physics by postulating an absolute standard of simultaneity, this is a controversial move, and undoubtedly a theoretical cost. What Bohm's theory delivers for this cost is a particularly straightforward continuity between microscopic causal explanations and macroscopic ones. At the microscopic level, explanations are given in terms of discrete objects—particles—following continuous trajectories according to a dynamical law. These explanations have the same form as the mechanical explanations we give at the macroscopic level; We explain, for example, why the window broke in terms of discrete middle-sized objects following continuous trajectories according to dynamical laws. Apart form the matter of nonlocality, Bohmian causal explanation is entirely familiar.

In the many-worlds theory, matters are not so straightforward: Causal explanations at the microscopic level are radically unlike those at the macroscopic level. At the macroscopic level we still explain things in terms of discrete objects following continuous trajectories, but at the microscopic level we explain things in terms of patterns in a wave. But note that there is nothing problematic about employing different kinds of explanation at different levels of description: As Wallace (2012, 48) points out, we don't explain tiger hunting patterns in terms of molecular dynamics. Both of the basic forms of explanation—in terms of discrete trajectories and in terms of waves—are perfectly acceptable. And the discrete trajectories of macroscopic objects (within a branch) can be shown to emerge as stable patterns in the underlying wave motion (Wallace, 2010). Many-worlds causal explanations, in this sense, conform perfectly well to the "classical ideal."

But nevertheless, many-worlds explanations have certain counterintuitive features. In particular, the causal stories we usually tell about particles passing through pieces of lab equipment will only sometimes reflect the actual underlying causal structure. Consider first a simple spin measurement performed on a spin-up "particle." The standard textbook causal story is that the particle traverses the apparatus, is deflected upward, and is detected when it produces a flash in the upper half of a fluorescent screen. The many-worlds story follows the standard story quite closely: A single wave packet traverses the apparatus in the way described, and hence the wave packet approximates the behavior of a particle.

But now consider the same spin measurement performed on a particle in a symmetric superposition state $(|\uparrow\rangle - |\downarrow\rangle)/\sqrt{2}$. Here the many-worlds causal story is that the initial wave packet splits in two, with one packet deflected upward and one deflected downward. This story is not the standard one: Particles, unlike wave packets, do not split in two.[8] Furthermore, the many-worlds story (so far) doesn't explain why I detected the particle in the upper-half of the screen, since the story is entirely symmetric between the result I observed and the one I did not. To break the symmetry, we have to bring in the additional explanatory machinery of self-location: I am located in one of the branches ensuing from this measurement process. Relative to my branch, just one wave packet is relevant to my result, and again we can regard the wave packet as approximating the behavior of a particle passing through the device. What is surprising is that self-location is a crucial element in causal explanation.

Finally, consider the case of two-slit interference discussed in Chapter 1. In this case there is no chance of explaining the outcome in terms of particles traveling through the slits to the screen: For each "particle" traversing the device, the wave must pass through both slits in order to generate the observed interference pattern. As in the prior example, the appearance of a flash at a determinate location of the screen can be understood in terms of the location of the observer in a particular branch: The wave hits the screen at many points at once, and each of these points generates a flash observed by one of my successors. But unlike the previous example, the provenance of this flash cannot be explained in terms of a particle-like object following a well-defined trajectory to that point. In particular, the explanation for the appearance of the flash at a particular point cannot be in terms of a wave packet that passes through one slit but not the other. Here the explanation of the passage of the "particle" through the device has to be fully reconceptualized in terms of waves.

So microphysical causation in many-worlds has two notable features: The replacement of particle trajectories with waves, and the ineliminable incorporation of the branch location of the observer in the explanation. But nevertheless, the many-worlds theory gives us perfectly well-defined and (potentially, at least) causally local explanations of the things we observe.

5.4 Collapses as Causes

What about spontaneous collapse theories? Like the many-worlds theory, there are no particle trajectories in theories like GRW, just wave motion, so many of the earlier comments about causation in many-worlds carry over to GRW. But the way in which the trajectories of macroscopic objects emerge from the underlying wave motion is very different. In particular, it is a physical collapse process, not an appeal to the branch location of the observer, that breaks the symmetry between observed and unobserved measurement outcomes.

In the case of a spin measurement on a superposition, for example, the wave packet splits into two, and this produces two nascent branches, but one of these branches is "snuffed out" at some point by the collapse process, so that only the causal story contained in the remaining branch is relevant to the observed result. In the case of two-slit interference, there are many such nascent branches, one for each point on the screen at which a flash might be observed, and again the collapse process "snuffs out" all but one.[9] In this case (as for many-worlds) no particular trajectory through the two-slit apparatus can account for the result we see; rather, a wave passes through both slits and hits the screen over a wide area. The collapse process picks out a particular point on the screen as relevant to the observed result, but not a particular path through the slits.

A couple of things about the GRW collapse process are worth noting. First, it has no classical analog: There is nothing in classical particle mechanics or classical wave theory that is anything like this discontinuous and fundamentally random *jump* between one wave distribution and a radically different one. This in itself is not problematic, but it does provide a sense in which spontaneous collapse theories have a more unfamiliar causal structure than either Bohm or many-worlds.

The second feature worth noting is that the GRW collapse process is not causally local. This, as we saw in the case of Bohm's theory, *is* a problem, albeit one that might be soluble. So let us take a look at how causal nonlocality arises in the GRW theory. Consider the situation shown in Figure 5.3 (c) at the end of the spin measurements performed on two entangled particles. Suppose that the result of the spin measurement on particle 2 is observed; this has the effect of correlating the location of particle 2 with the locations of a large number of other particles, for example, in the brain of the observer. This in turn makes the chance of a GRW collapse very high over a very short period of time. The collapse will be centered on one or other wave packet, with probability 1/2 each. Suppose it is centered on the top-left packet; the result is that the amplitude of this packet becomes almost 1 and the amplitude of the bottom-right packet becomes almost 0.

But note that each of these packets represents the locations of *both particles*. According to the GRW criterion for property possession described in Chapter 4, the collapse event results in each particle acquiring a determinate location, since its wave amplitude is now almost all contained in the top-left packet. That is, the measurement on particle 2 instantly causes particle 2 to acquire a determinate position, and it *also* instantly causes particle 1 to acquire a determinate position, no matter how far apart the two particles are.[10] So the GRW theory, like Bohm's theory, violates local causation and hence conflicts with special relativity.

As in the case of Bohm's theory, one response to this problem is to propose that special relativity, like quantum mechanics, is incomplete, and must be completed by the addition of an absolute standard of simultaneity. The costs are as before: While nonlocal causation is not in itself conceptually problematic, the required modification of special relativity compounds the already ad hoc flavor of the GRW theory. So is it possible to construct a spontaneous collapse theory that does not conflict with special relativity? Maybe. Two distinct strategies have been pursued in this direction. The first is to construct a spontaneous-collapse analog of the retrocausal hidden variable approach. The second is to take advantage of the lack of determinate particle locations for microscopic systems according to spontaneous collapse theories. Let us take them in turn.

So suppose we try to rescue locality in the causal explanation of measurement results by violating Bell's independence assumption instead of his locality assumption. In the case of spontaneous collapse theories, this means that the collapse process is affected not only by what has happened to the system in the past but also by the measurements that will be performed on the system in the future. The transactional interpretation devised by Cramer (1986) and elaborated by Kastner (2013) is the best-developed example of this approach. The basic mechanism in the case of spin

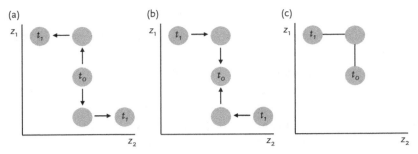

FIGURE 5.6 Entangled particles: the transactional interpretation.

measurements performed on entangled particles is shown in Figure 5.6. As in standard GRW, waves travel out from the particle emission event at time t_0 to the measuring devices, where they are deflected up or down according to spin, and then detected, for example, by a fluorescent screen at time t_1. The causal development of these "offer" waves is shown in Figure 5.6 (a). Unlike standard GRW, the detection of the arriving waves does not itself precipitate collapse, but instead triggers the emission of "confirmation" waves of the same amplitude that travel backward in time from the detection events at t_1 to the emission event at t_0, essentially retracing the paths of the original "offer" waves. The causal development of the confirmation waves is shown in Figure 5.6 (b): Note that the arrows in the diagram indicate the direction of *causation*, not the direction of time. It is the arrival of the confirmation waves at the source that triggers the collapse event: One offer-confirmation pair is chosen at random as actual, with probabilities given by the squared amplitudes of the confirmation waves returning to the source. The completed "transaction" between the source and the detectors is shown in Figure 5.6 (c): Note that since it represents a wave in a six-dimensional space, the completed transaction represents *two* particles, one detected as spin-up and the other detected as spin-down.

The retrocausal aspect of this approach means that there is no need for a violation of locality to explain the Bell correlations: The properties of the particles at the source are influenced by waves returning from the detectors, and hence can reflect the measurements that will be performed on them, as explained earlier for hidden variable theories. However, a collapse theory does not lend itself naturally to a retrocausal treatment. Retrocausal theories are typically understood in a temporal "block universe"—a picture in which future events are just as real and determinate as present events—so that future events can act as causes (Price, 1994, 315). But then in the case of the transactional interpretation, at the beginning of the experiment there is already a fact of the matter about which offer-confirmation pair is actual, so the causal story at the heart of the theory cannot be taken literally: At no time is there a multiplicity of confirmation waves returning from future measurement events, so such waves cannot trigger collapse. Cramer (1986, 661) suggests that the causal story should *not* be taken literally, but then Maudlin (1994, 198) complains that without it, the transactional interpretation has no explanatory content. Kastner (2013, 67) attempts to take the causal story (more or less) literally by adopting a dynamical view of time, according to which the nonactualized confirmation waves are "real possibilities" that fall away as time progresses, so that the collapse process is constitutive of the passage of time. Evaluation of this proposal would take us too far into the philosophy of time, but it is clear that however we should conceive of the transactional interpretation, if it yields coherent causal explanations, they deviate significantly from the classical ideal, not least because of the dependence of the present on the future.[11]

The second strategy for constructing a spontaneous collapse theory that is consistent with special relativity trades on the fact that property possession in spontaneous collapse theories can be a discontinuous, intermittent kind of thing; if particles only have determinate locations after collapse events, then perhaps systems simply do not have enough location properties at any instant of time to constitute a violation of locality. This is the "flashy" strategy devised by Tumulka (2006a, 2006b) and discussed briefly in Chapter 4. Tumulka takes his lead from some observations of Bell. As Bell notes, although the GRW collapse process as a whole is spatially spread out, the *center* of a collapse event picks out a precise location in space and time—a space-time point. Bell suggests that these events should be taken as the ontological heart of the collapse theory, and that "a piece of matter is a galaxy of such events" (2004, 205). Tumulka adopts this ontology of "flashes" at precise space-time points and builds a dynamical theory that generates the patterns of flashes in a manner consistent with special relativity. The central trick is that the probability distribution for these flashes for a particle is not defined over the $t = 0$ surface ("the whole of space right now"), since special relativity tells us that this surface depends on an arbitrary choice of a standard of simultaneity. Rather it is defined over a space-time surface that is invariant under different choices of such a standard.[12]

This deals with collapses for a single particle, but it doesn't deal with correlations between particles. Here Tumulka (2006b, 350) admits that his current model is nonlocal, but echoing earlier work by Myrvold (2002), he suggests that the conflict with relativity can be mitigated by making causal stories depend on your point of view. Suppose both particle 1 and particle 2 have their spins measured. According to some choices of coordinates, the measurement on particle 1 occurs first, and this precipitates the collapse that leads to particle 2 having a determinate location. According to other choices of coordinates, the measurement on particle 2 occurs first, and this precipitates the collapse that leads to particle 1 having a determinate location. Special relativity entails that the choice between the two sets of coordinates is entirely conventional—so the suggestion here is that the direction of causation in this example is a matter of convention, too. Whether it is coherent to relativize causal explanation to a set of coordinates in this way is not a trivial question. But clearly Tumulka's suggestion is not a defense of *locality* for collapse theories: Whatever set of coordinates you choose, the causation involved is faster than light, and for one choice of coordinates it is instantaneous.[13]

So Tumulka's proposal is revisionary with respect to causation in at least two respects. First, the direction of causation, and hence the details of the causal explanation, depends on the conventional choice of a standard of simultaneity. There is no perspective-neutral causal story. Second, since flashes are discrete space-time events, causation at the level of the flashes is intermittent rather than continuous. A flash here and now can cause a flash over there later without any intermediate causal links. In fact, the causal elements in flashy GRW are very sparse. A macroscopic object is a galaxy of flashes, but for a microscopic system, it is very unlikely that there will be any flash at all within a humanly reasonable time frame. So flashy GRW is rather nihilistic when it comes to micro-ontology—for most microscopic systems there is nothing there at all! The causal account of the behavior of such systems is packed into the flashes that occur in the macroscopic objects involved in the preparation and detection of this system, with the intermediate "system" itself not existing as a concrete causal process at all.

5.5 Conclusion

Bohr is almost certainly wrong when he says that quantum mechanics requires us to renounce causation at the microscopic level. Causation is a varied and flexible notion, and it is hard to imagine a physical theory that isn't causal in any sense. All of the theories canvassed here provide causal explanations of quantum phenomena. Nevertheless, if we were looking for something that resembles familiar classical causal explanation, we don't necessarily find it in quantum mechanics; the accounts given here modify the structure of causal explanations in various interesting ways.

When we observe a flash in the upper half of a fluorescent screen and take it to indicate that the measured particle is spin-up, we typically picture the particle following a continuous trajectory through the measuring device to the screen. Bohm's theory retains this picture, but at the cost of admitting instantaneous causation at a distance. Retrocausal hidden variable theories promise to recover causal locality, but at the cost of admitting causes that occur after their effects. The many-worlds theory jettisons particle trajectories and in general requires a reconceptualization of the underlying causal explanation of microscopic processes in terms of waves. These wave explanations equally entail both observed and unobserved outcomes, so that an appeal to the location of the observer in the wave structure has to be added as part of the explanation. The GRW theory also requires explanation in terms of waves, but it replaces the appeal to self-location with an appeal to a discontinuous, probabilistic jump in the wave state. This jump, like Bohm's particle dynamics, incorporates instantaneous causation at a distance. Strategies for recovering causal locality either allow that causes can occur after their effects, or that whether an event is a cause or an effect can be a matter of convention.

The whole situation regarding causation in quantum mechanics is quite complicated. But the only feature of causal explanations considered here that is straightforwardly problematic is nonlocality. The causal stories told by Bohm's theory and the GRW theory conflict with special relativity and require us to modify special relativity by adding an (undetectable) standard of absolute simultaneity. Among the three major interpretations, the many-worlds theory is at a distinct advantage on this score.

6| Determinism

ARGUABLY THE MOST COMMON metaphysical consequence claimed of quantum mechanics is that it is indeterministic—that probabilities enter the theory not as a matter of our ignorance, but because there are genuine, irreducibly probabilistic processes in the world. It's easy to see how one might come to such a conclusion based on the central place of the measurement postulate in the standard theory: The measurement postulate tells us that the state "collapses" to an eigenstate on measurement, where the particular eigenstate chosen is a matter of chance rather than deterministic law. However, since the measurement postulate is untenable, it cannot be the basis of any conclusions about determinism.

If we look instead to the versions of quantum mechanics that have arisen in response to the failings of the measurement postulate, the situation is far more equivocal. As we saw in Chapter 3, the GRW theory is fundamentally indeterministic: It postulates a new dynamical law that incorporates irreducible probabilities in two separate places. But Bohm's theory is entirely deterministic: The additional dynamical law involves no probabilities, and uncertainty over measurement outcomes only enters as a result of our ignorance of the precise locations of the Bohmian particles. So this is an area where underdetermination has bite: On some interpretations quantum mechanics is fundamentally indeterministic, and on some interpretations it is entirely deterministic. At present we cannot make any definite pronouncement concerning indeterminism in the world based on quantum mechanics.

But even though quantum mechanics doesn't provide a definitive answer, it reshapes the debate over determinism in interesting ways. In large part, this reshaping comes as a result of thinking about the status of determinism in the many-worlds theory. If the many-worlds theory is true, is the world deterministic or not? More precisely, how (if at all) do the probabilities of the Born rule arise in a theory in which every possible outcome of a measurement definitely occurs? Thinking about such matters gives us new ways to think about determinism, uncertainty, and the foundations of probability; these considerations take up the bulk of this chapter. At the end of the chapter, because such questions often appear in more popular presentations of quantum mechanics, I also take a look at the relationship between the kinds of determinism and indeterminism posited by quantum theories and the vexed question of free will.

6.1 Uncertainty

So what does the many-worlds theory tell us about determinism? Superficially, the theory is entirely deterministic: The fundamental dynamical law of the theory, the Schrödinger equation, is a deterministic law, and there are no probabilistic collapses. But there is another sense in which the many-worlds theory looks indeterministic. Suppose you are an observer about to perform a z-spin measurement on a particle whose state is a symmetric superposition of z-spin up and z-spin down eigenstates. What should you expect to see? In Bohm's theory, while you don't know what you will see, there is a fact of the matter about what you will see dictated by the positions of the Bohmian particles right now.[1] But in many-worlds, there are no such particles, and hence (it seems), there is no such fact. Rather, the state simply does not determine what you will see. From this first-person perspective, the many-worlds theory seems indeterministic.

As a first pass, then, we might say that the many-worlds theory is objectively deterministic but subjectively indeterministic: Even though the global state of the world evolves according to a deterministic law, the experiences of an observer within a branch are necessarily unpredictable in advance. We can couch this in terms of uncertainty: If you know the initial physical state of a system, you cannot be uncertain of its final physical state, but you can be uncertain about your own later experience.

But this is immediately problematic. A fairly common supervenience claim about the relationship between the mental and the physical is that the physical state of a system that includes a conscious observer determines the mental state of that observer. The "determination" here isn't dynamical in nature, but is rather a matter of which physical states underlie which mental states: Maybe different physical states could underlie a single mental state, but a single physical state couldn't underlie two different mental states. Now consider the z-spin measurement again: If the premeasurement physical state determines the postmeasurement physical state (in the ordinary dynamical sense), and the postmeasurement physical state determines your mental state (in the nondynamical supervenience sense), then given the premeasurement physical state, your postmeasurement mental state couldn't be other than what it is.

Let us spell that out a little bit. Before the measurement, you know exactly what will happen, from a physical point of view: The current state will evolve to a state with two branches, a spin-up branch and a spin-down branch. And because you know that the physical determines the mental, you know exactly what will happen from a mental point of view, too: Your current mental state will evolve into two distinct mental states, one experiencing the spin-up outcome and one experiencing the spin-down outcome. That is, you know that you will have two distinct successors, each of whom is connected to you in the usual way via memory and continuity of experience, but who have very different current experiences. In such a situation, you can't wonder which outcome you will see, since you know that you will have one successor seeing the spin-up outcome and one seeing the spin-down outcome. Neither can you wonder which of the successors is really *you*, since they each have an equally good claim to be you: You will equally become each of them. There is nowhere for uncertainty to get a foothold.

Of course, a many-worlds observer *might* be uncertain about the quantum state, but there is no reason she *has* to be, and since the quantum state represents all the physically relevant facts according to the many-worlds theory, there in nothing she need be uncertain of. This is a different situation from Bohm's theory: Here an observer is *necessarily* uncertain of the precise location of the Bohmian particles within the wave function, since this uncertainty is a provable consequence of the theory (Bohm, 1952, 182). But without uncertainty in the many-worlds theory, it looks like there can be no place for probability either. The many-worlds theory is objectively deterministic, so if there are probabilities here, they must be subjective—they must quantify uncertainty. And without probability, the many-worlds theory fails. Quantum mechanics makes its predictions in the form of probabilities, and the success of quantum mechanics lies in the agreement between these probabilities and the relative frequencies of the outcomes we observe. If the many-worlds theory has no place for probabilities (other than zero and one), then it is an empirically unsuccessful theory.

Is there any way we can find the requisite uncertainty in the many-worlds picture? It is not hard to *generate* some uncertainty. For example, suppose you close your eyes during the spin measurement. After the measurement, but before you open your eyes, you (that is, each of your successors) can be legitimately uncertain about whether you will see spin-up or spin-down (Vaidman, 1998). But by itself this doesn't get us very far. For one thing, this uncertainty isn't an essential feature of the worlds: You could eliminate the uncertainty by keeping your eyes open.[2] For another, the uncertainty is strictly limited to the postmeasurement period. Although, before the measurement, you know that your successors will be uncertain about the result they will see when they open their eyes, that doesn't mean that there is anything for you to be uncertain of right now. Before the measurement, you know where in the branching structure you are, and where in the branching structure each of your successors is; it is only after the measurement that one of your successors can legitimately wonder where in the branching structure they are.[3] If we want uncertainty to ground probabilistic prediction, then strictly postmeasurement uncertainty won't do.

But for you to be uncertain before the measurement about what you will see later, it seems that there would have to be some unique fact about what you will see later—spin-up rather than spin-down, say. Many-worlds quantum mechanics, together with the earlier supervenience of the mental on the physical, precludes the existence of such a fact: The spin-up branch and the spin-down branch are equally real, and supervenience ensures they contain the experience of seeing a spin-up result and seeing a spin-down result, respectively. So one might be tempted to give up on the supervenience claim at this point, and say that one of the postmeasurement branches contains a conscious experience and the other does not, despite the fact that a state physically just like the nonconscious branch might in other circumstances (e.g., on other runs of the experiment) contain a conscious experience.

This is the so-called *single-mind* theory of Albert and Loewer (1988, 205). It is not particularly attractive, as Albert and Loewer are quick to point out. Rejecting the supervenience of the mental on the physical looks like a fairly desperate move, resulting in a rather strong kind of dualism. One term in the superposition is associated with a mind and the other is not, even though there is no salient physical difference between them. Indeed, given a functional view of mental activity, it looks like the mere stipulation that one branch contains a mind and the other does not is bound to fail, since both branches contain the physical brain processes that we associate with mental activity. Furthermore, we need to postulate a new law that generates the probabilities that the premeasurement mind will end up associated with each of the two postmeasurement outcomes. The motivation for the many-worlds theory is to avoid altering the physics of standard quantum mechanics, and now it looks like we are forced to do just that. The only defense might be an insistence that the minds are not *physical* entities, so the law in question wouldn't be a *physical* law, but I have little idea what such a distinction might mean.

So suppose instead we postulate that each conscious brain is associated with *many* minds rather than a single-mind, where some of the minds follow each branch, and the proportion of minds associated with a branch is given by the squared wave function amplitude. Here there is a fact of the matter in advance about which result each mind will experience, so before the measurement you can be uncertain about whether your mind is one of those that will see spin-up or one of those that will see spin-down. Since the squared wave function amplitude of a branch is a real number, and since the proportion of minds associated with this branch has to match this number, we will need a continuous infinity of minds associated with each physical body. So whereas under the single-mind theory each body (in a branch) is associated with at most one mind, here each body (in a branch) is associated with an infinite number of minds.

This is Albert and Loewer's *many-minds* theory (1988, 206). At first glance, it seems even more desperate than the single-mind theory: If postulating a single nonphysical mind per body is unacceptable, postulating an infinite number of such minds per body is surely beyond the pale. But there is a sense in which the many-minds theory is an improvement over the single-mind theory. Note that since every branch containing a functioning brain is associated with minds, the functional view of mental activity can be retained. And note that, given the ontology of many-worlds quantum mechanics, many minds can supervene on the physical state even though a single mind can't: The state determines a unique distribution of minds over physical branches. But on the other hand, the individuation of the minds is not given by the physics: The minds associated with a given branch are physically indistinguishable, so if they are distinct individuals, this individuation must be irreducibly nonphysical. The specter of dualism looms again. Put another way, the fact that the number of minds associated with a branch is given by its squared amplitude must be a brute *postulate* about the relationship of the mental to the physical, and again we are forced to *add* something (and something nonphysical) to standard quantum mechanics.

Is there any way we can individuate the minds associated with a branch without appeal to irreducible nonphysical entities? There is nothing in the state at a time that could underwrite this individuation—but the evolution of the state *over* time might do the trick. That is, the *subsequent* branching of the current branch into a multitude of distinct sub-branches could make it the case that there are a number of distinct individuals corresponding to the current branch. This is a familiar move. Parfit (1971) argued that in cases where a single individual splits into two copies the concept of personal identity becomes inapplicable; David Lewis (1976) resisted this argument by proposing that a person is a temporally extended entity consisting of a complete life history. According to the latter, there are two people present both before and after the splitting event; before the splitting event their histories coincide, and afterward they diverge.

This Lewisian ontology of people (or minds) looks like just what we need here. Personal splitting events are of course endemic in many-worlds quantum mechanics. Associating people with complete personal histories yields the multiple premeasurement minds of the many-minds theory, but without the need to postulate anything over and above the branching quantum state. This approach to understanding many-worlds ontology has been defended by Saunders and Wallace (2008) and Wallace (2012, 283). They claim in particular that it gives us an account of genuine uncertainty in the many-worlds theory, since as in the many-minds theory, prior to a measurement you can be uncertain about which of the premeasurement people you are, and hence uncertain about which outcome you will see.

But this potential source of uncertainty needs to be treated circumspectly. Consider again the simple spin measurement. In that case, the premeasurement people fall into two classes—the ones that will see spin-up and the ones that will see spin-down. Prima facie, then, each premeasurement person can be uncertain about which class they belong to. But how should this uncertainty be analyzed? The obvious story is that for each person, there is a fact of the matter about whether "I will see spin-up" is true, and the person is ignorant of this fact. The difficulty with this analysis concerns the referent of "I." For the analysis to work, when all the co-located premeasurement people think "I will see spin-up," each of them must succeed at referring to just that single person and not to all the others. What could ground this success?

To use a well-worn analogy, suppose two roads, US 27 and FL 80, overlap right here but diverge 10 miles ahead. Can I use the expression "this road" to refer to US 27 but not FL 80? Can I be uncertain about whether *this road* goes to Sebring?[4] It is hard to see how. Even though there are two distinct roads stretching across Florida, there is only one road *segment* right here (or road *stage* in Sider's [1996] terminology). My indexical "this road" determinately picks out this road segment—after all, I can point to it—but does not determinately pick out one road or the other. Similarly in the many-worlds case, even though there are many distinct people stretching across time, there is only one person segment (or person stage) right now. My indexical "I" determinately picks out this person segment—after all, I can point to it—but it does not determinately pick out one extended person over the others (Tappenden, 2008). Without determinate self-reference, there is no way for me to express a proposition that I can be uncertain of.

Perhaps this worry can be defused. Ismael (2003) argues that although the proposition I am uncertain about is inexpressible in advance, this is just a linguistic peculiarity with no ontological consequences: There is a *fact* about what outcome I will see, even if I cannot express the fact in advance. But this is not so clear; the "fact" involves an essential indexical component, and it is arguable that there are no facts in advance about what outcome I will see, expressible or otherwise (Lewis 2007b). Sider (2014) attempts to secure the relevant facts by postulating a new kind of indexical reference that points directly to one of the postmeasurement individuals, but no particular one of them. Granted this novel kind of indeterminate reference, one can arguably generate some kind of premeasurement uncertainty: Whether *I* (i.e., a generic postmeasurement successor) will be among those successors seeing spin-up. But it is not so clear whether this is the *right* kind of uncertainty; it is not so clear that I am wondering about a *generic* successor.

So even if an appeal to Lewisian four-dimensional people can dispel some of the worries we might have about personal identity in branching worlds, it does not seem to recover premeasurement uncertainty about the outcomes of measurement. But maybe this just indicates that we were looking for something too familiar—something too close to our ordinary conception of uncertainty. Ordinarily, if you are uncertain about something, there must be some fact that you are ignorant of—something that, were you to learn it, would eliminate the uncertainty and hence improve your epistemic situation. I can be uncertain whether it is 31°C or 32°C in Miami today, but I can't be uncertain whether it is 31.999°C or 32.000°C, because there are just no facts about atmospheric temperature over the relevant region of space (and time) at that degree of accuracy.

Now consider how things go in the many-worlds case. In a simple spin measurement, for you to be uncertain about whether you will see spin-up or spin-down, there must be some fact about which result you will see that you are ignorant of. The trouble, as we have seen, is that it is hard to find such a fact in many-worlds quantum mechanics. But this is not because the claim "I will see spin-up" is too fine-grained, as in the temperature case; rather, there is no *unique* fact because there are too many such facts. It is a fact that you (or, rather, some successor of you) will see spin-up, *and* it is a fact that you (some successor of you) will see spin-down. Some of your successors will learn one of these facts, and the rest will learn the other. But the key point is that each of your successors will have her epistemic situation improved when she learns her respective branch-relative fact, even if that improvement in epistemic situation can't be analyzed in terms of some unique preexisting fact. So while uncertainty in the usual sense about future measurement outcomes doesn't exist in many-worlds universes, a close analog does. The question is whether this pseudo-uncertainty is enough for present purposes.

6.2 Probability

The present purpose, of course, is to recover probability in the many-worlds theory. The situation so far is this. By individuating people based on their whole histories, we can identify multiple people existing even prior to a measurement. Maybe this means that each such person can be genuinely uncertain about future measurement results, although this remains unclear. But in any case, each such person can be pseudo-uncertain in the aforementioned sense about future measurement results. This is a somewhat promising start, but we need to assure ourselves of two further things. First, we need to make sure that the appropriate measure of this pseudo-uncertainty obeys the probability axioms—that it can be properly characterized as probability. Second, we need to make sure that this probability measure assigns the same probabilities to measurement outcomes as the Born rule. That is, for the many-worlds theory to be empirically adequate, it must not only assign probabilities to outcomes; it must assign the *right* probabilities.

The first thing that might strike you is that if the people are distributed in the right way over the various branches, then the problem solves itself. That is, you might think that if the number of people in each branch is proportional to the squared amplitude of the branch, then clearly each person should have a probability distribution over future measurement results given by the Born rule. Unfortunately, though, Lewisian people will not in general be distributed according to the Born rule. This is because the number of Lewisian people in a branch has to do with the subsequent branching behavior of that branch, not its squared amplitude. Two branches with equal squared amplitude can differ in their subsequent branching structure, so that one branch contains more Lewisian people than the other. And two branches with different squared amplitudes can have the same subsequent branching structure, and hence contain the same number of Lewisian people.[5]

So we need another approach. The oldest and least successful strategy approaches probability in the many-worlds theory via the relative frequency of measurement outcomes in branches. Suppose we have a number of particles all prepared in the same z-spin superposition state $a|\uparrow\rangle + b|\downarrow\rangle$, and we measure the spin of each particle successively. What sequence of results will we obtain? Well, every possible sequence will be exemplified on some branch or other. But as the number of measurements tends to infinity, one can prove that the total squared amplitude of branches in which the frequency of spin-up results is close to $|a|^2$ tends to 1.[6] This is hopeful: It suggests that one will almost certainly see Born-rule frequencies in the long run. But the hope is short-lived; the "almost certainly" here depends on the identification of squared amplitude with probability, which is precisely the question at issue. So (as first pointed out by DeWitt, 1970), the argument is circular. Indeed, if we ask instead about the *proportion* of branches in which the frequency of spin-up results is close to $|a|^2$ (the proportion of the *number* of branches rather than the total squared amplitude), one can easily prove that this tends to 0 as the number of measurements tends to infinity.[7] Because the number of Lewisian people follows the number of branches rather than their squared amplitude, one might conclude that the Born rule is not only unmotivated in the many-worlds theory but wrong.

The newest and most successful strategy for locating probability in the many-worlds theory approaches the question subjectively, rather than in terms of frequencies (Deutsch, 1999; Wallace 2003b, 2007, 2012). That seems appropriate: Because the many-worlds theory is physically deterministic, it looks like the source of any probability here must be in terms of the beliefs of branching subjects. The basic approach is to impose certain intuitive rational constraints on the preferences of agents, and then to prove that any agent who satisfies those constraints will have degrees of belief satisfying the Born rule.

Some of the constraints are fairly standard stuff in decision theory. Suppose that an agent has a choice among various acts, where an act involves performing a particular quantum measurement. Acts generally induce branching, and it is assumed that the agent receives certain specified rewards on the resulting branches. Presumably the agent's preferences over these acts need to satisfy certain constraints if they are to be rational. For instance, Wallace (2012, 167) assumes that rational agents' preferences satisfy *ordering* and *diachronic consistency*. Ordering is the assumption that the agent's preferences are transitive, reflexive, and connected, and diachronic consistency is the assumption that an agent's preferences do not conflict with those of her future selves. More precisely, let U, V, and W be any acts, and let $U \succeq V$ mean that the agent either prefers U to V or is indifferent between them. Then we can state these assumptions as the following axioms:

Ordering: An agent's preferences form a total ordering, in the sense that $\succeq$ is transitive (if $U \succeq V$ and $V \succeq W$ then $U \succeq W$), reflexive ($U \succeq U$), and connected (either $U \succeq V$ or $V \succeq U$).

Diachronic consistency: Suppose act U is followed by branching into i successors, and each successor has a choice between act V_i or act V'_i. If $V_i \succeq V'_i$ for each of the i successors, the agent's current self prefers U followed by the V_i to U followed by the V'_i, or is indifferent between them.

The rest of the constraints are less familiar, and specific to the many-worlds context. Again, I follow Wallace's axiomatization (2012, 170). Wallace assumes the following:

Macrostate indifference: The agent cares only about the macrostate, not about which microstate instantiates it.

Branching indifference: The agent doesn't care about branching per se.

State supervenience: The agent's preferences depend only on what physical state they actually leave her branch in.

Solution continuity: Sufficiently small changes to the acts do not affect the agent's preferences between them.

Given these assumptions, Wallace proves that the agent's preferences can be represented by a unique utility function over the set of rewards (2012, 172). The utility function represents the agent's preferences in the following way. Each act generates a set of branches containing rewards, and one can obtain a number by multiplying the utility of each reward by the squared amplitude of the branch in which the reward occurs and adding the results. The agent prefers acts with a bigger number to acts with a smaller number. That is, the numbers function as expected utilities in generating the agent's preferences—the agent maximizes expected utility—and the squared branch amplitudes function as probabilities in constructing those expected utilities. The crucial point is that rationality dictates that the expected utility calculation uses *squared branch amplitude* in the "probability" slots, and hence vindicates the use of the Born rule to generate the probabilities of outcomes.

The proof is somewhat involved, so I will not reproduce it here. But it is worth noting that nothing in the proof hinges on the existence of uncertainty in the many-worlds theory. As Greaves (2004) has pointed out, one could just as well take squared branch amplitude as a measure of how much a rational agent *cares* about each of her successors, rather than a measure of any kind of uncertainty regarding her successors. Wallace takes a functional view of probability, so since squared branch amplitudes function like probabilities, they *are* probabilities, whether or not they are connected to uncertainty (2012, 148).

The proof is unimpeachable, but the conclusion is somewhat surprising. Similar proofs in the foundations of probability (e.g., Savage, 1954) can be used to show that if an agent's preferences obey certain intuitive constraints, then those preferences can be uniquely represented by a utility function and a probability distribution over outcomes, such that the agent's preferences maximize expected utility. But the difference is that while this latter proof shows that the agent's preferences are representable by *some* probability function, it is silent on the nature of the probability function. This seems exactly right: While it may well be a matter of rationality alone that your degrees of belief obey the probability axioms, it is surely not a matter of rationality alone that you have any particular set of degrees of belief. And yet the Deutsch-Wallace proof entails that the agent's degrees of belief should not only obey the probability axioms but also should conform to the Born rule; they should be the squared amplitudes of the branch weights.

The reason that this more restrictive conclusion follows, of course, is the second set of constraints—those that are specific to the many-worlds context—and one might reasonably suspect that some of them may not be a matter of rationality alone. In particular, branching indifference has received a good deal of critical attention in the literature (e.g., Albert, 2010; Kent, 2010; Lewis, 2010; Price, 2010). It is not hard to see why: Given the branching structure of people in many-worlds, it seems that it is rationally incumbent on an agent to *pay attention* to branching, not to ignore it. For example, suppose that a given act causes you to branch into four copies, three of whom receive reward A and one of whom receives reward B. Intuitively, it looks like you should give reward A three times the weight of reward B in calculating your expected utility, irrespective of the squared amplitudes of the branches. After all, from a subjective point of view, the squared amplitude of the branch you occupy is irrelevant; a high-amplitude branch feels just like a low-amplitude branch. On the other hand, how many of your successors will experience a given reward seems to be a clear matter of concern. So intuitively, it looks like the expected value of a reward should be independent of the squared amplitude of the branches it appears in, and determined entirely by how many branches it appears in, contrary to branching indifference.

Wallace defends branching indifference on the grounds that there is no number of branches. We have been assuming in the presentation so far that the number of branches in a given situation can simply be read off the quantum state. For example, I just described an observer branching into four copies. This looks pretty straightforward: We could just have the observer measure the spins of two spin-1/2 particles in superposition states in succession, generating four terms in the state and hence four branches. But in reality things are much more complicated. Macroscopic objects like people undergo branching events all the time, many times per second, based on quantum interactions with their environments. So in reality, there are not four branches, but many. How many? Well, it depends on choices we make about how to write down the quantum state—what coordinate system to use to define the vector space—and when to count two terms as separate branches. These choices are purely conventional, but they can make a huge difference—orders of magnitude—in how many branches we say there are. So there is no fact of the matter, not even an approximate one, concerning how many branches are associated with a given outcome. If there is no fact of the matter, then presumably it makes no sense to base one's decisions on how many branches are associated with an outcome.

However, you might suspect that even if there is no objective *count* of the number of branches, there is an appropriate measure we can substitute instead. For example, you might think that the *proportion* of branches associated with a particular macroscopic outcome provides the appropriate measure. Initially, this looks promising. If we perform a spin measurement on a particle whose state is $a|\uparrow\rangle + b|\downarrow\rangle$, then the number of branches associated with spin-up and spin-down may be radically dependent on our representational choices, but there is no reason to think that the proportion of branches associated with spin-up and spin-down will be sensitive to our choices in this way. Whatever choice we make will presumably affect the branch count for spin-up and spin-down outcomes in the same way, since they consist of the same macroscopic objects in the same environments, so the proportion of branches associated with spin-up and spin-down outcomes will be practically independent of our choice.

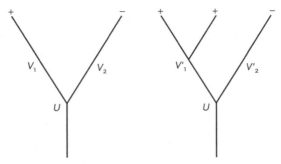

FIGURE 6.1 Two branching scenarios.

But the move from number of branches to proportion of branches leads to the violation of another of Wallace's axioms, namely diachronic consistency. Consider the two branching scenarios depicted in Figure 6.1. In each scenario, act U produces two successors, one of whom will receive a positive outcome and the other of whom will receive a negative outcome; U might be a spin measurement on a particle whose state is $a|\uparrow\rangle + b|\downarrow\rangle$. Your successors are told which outcome they will receive, and the positive outcome successor is given the choice of branching into two further copies (act V'_I on the right) or not (act V_I on the left). That is, act V'_I is a second spin measurement performed on a particle in a similar state, and V_I is an act that does nothing. The negative outcome successor is given no such choice, but we can represent that as the choice between two acts V_2 and V'_2 that each do nothing. Acts U and V'_I each double the number of branches at the relevant points in the branching structure, and acts V_I, V_2, and V'_2 leave the number of branches unchanged. In addition, there is branching going on in the background, which presumably goes on at a roughly constant rate whatever else we do. That is, however we choose to count branches, in the left-hand scenario the two lines at the top represent half of the branches each, and in the right-hand scenario the three lines at the top represent one third of the branches each.

If your preferences follow the proportion of branches containing a positive outcome, then clearly you prefer act U followed by V'_I and V'_2 to U followed by V_I and V_2. In Figure 6.1, you prefer the right-hand scenario to the left-hand scenario because two thirds of your successors receive the positive outcome in the right-hand scenario, and only half of your successors receive the positive outcome in the left-hand scenario. But now consider the choice between V_I and V'_I faced by your positive outcome successors. The proportion of *their* successors who receive the positive reward is *one*, whether they choose to branch or not, so presumably they are indifferent between

V_1 and V'_1. Furthermore, your negative outcome successors are clearly indifferent between V_2 and V'_2, since both of these acts do nothing. But this is a straightforward violation of diachronic consistency: Your successors are indifferent between acts V_i and V'_i, and yet you prefer U followed by the V'_i to U followed by the V_i.

Diachronic consistency is a plausible axiom of standard decision theory, not one of the special many-worlds axioms, so violating it is not an attractive option. So it seems that we really do need a measure of the *number* of branches associated with an outcome, not just the proportion. But even so, it looks like the part of that number that is a matter of convention can be factored out of decision problems. Consider again the choice depicted in Figure 6.1. The actual number of branches represented by each line at the top of the diagram will depend sensitively on the choices we make about how to individuate branches. But whatever choices we make, given the branching structure outlined here, each of the five lines at the top of the diagram will represent roughly the *same* number of branches. This gives your positive-outcome successor a clear rationale for choosing further branching: *Twice as many* successors receive the positive outcome. Hence, the violation of diachronic consistency is defused: You prefer U followed by the V'_i to U followed by the V_i precisely because your positive outcome successor prefers V'_1 to V_1. So even if branches can't be counted precisely, or even approximately, they can be counted up to an irrelevant, conventional scaling factor that affects all choices equally.

Crucially, there is no reason to think that the branch count produced in this way will depend on the squared amplitudes of the branches, since the branching is a result of interactions—measurements and background activity—that affect high-amplitude and low-amplitude branches equally. In the example, the amplitudes a and b in the state used to generate branching at U and V' play no role in the subsequent analysis. This suggests that a rational agent should treat high-amplitude and low-amplitude outcomes as of equal weight for decision purposes, in violation of the Born rule. Of course, this judgment by the agent could be mistaken—perhaps there is an unforeseen source of further branching that affects one branch but not the other—but a decision rule doesn't have to be infallible, just generally reliable.

So the Deutsch-Wallace decision-theoretic proof that there are probabilities in many-worlds quantum mechanics is promising, but it can be challenged on the basis that one of their assumptions, branching indifference, is not an axiom of rationality. Furthermore, the reasoning that undermines branching indifference seems also to promote a rival probability measure to the Born rule—namely that given by the number of branches associated with an outcome (up to an irrelevant scale factor). If this rival rule is indeed a consequence of the many-worlds approach, then many-worlds quantum mechanics is simply empirically inadequate: The many-worlds approach stands or falls with the success of arguments like those of Deutsch and Wallace.

6.3 Immortality

Another way to gain insight into probability in many-worlds quantum mechanics—and a fun topic in its own right—is to think about what it entails about death. So consider the following game of quantum Russian roulette. You perform a spin measurement on the state $a|\uparrow\rangle + b|\downarrow\rangle$. The measuring device is wired to a bomb in such a way that if the outcome is spin-up, nothing happens, but if the outcome is spin-down the whole lab, including you, is instantly vaporized. After the measurement, you have at least one living successor seeing spin-up (and presumably many on a realistic account of branching). But having one living successor is all that it takes to survive. Hence, you will survive—no matter how small the squared amplitude of the spin-up branch. Furthermore, any potential cause of death is like this. Suppose instead that you put an ordinary gun to your head. Since every physically possible outcome is instantiated in some branch according to the many-worlds theory, and since it is physically possible (although classically highly unlikely) that the bullet will spontaneously vaporize in the barrel, there is a (very low-amplitude) term in the quantum state in which you have a living successor. The same goes for cancer, heart attacks, and falling off tall buildings; there is always a small term in which you survive.

On this basis, several commentators have claimed that the many-worlds theory entails that everyone is immortal (D. Lewis, 2004; P. Lewis, 2000; Price, 1996; Tegmark, 1998). Certainly the many-worlds theory has the following consequence: At any future time, there is a branch containing a living successor of you. The question is whether this is sufficient for immortality. The reason for doubt is that if the attempt to accommodate probabilities in many-worlds succeeds, then your rational expectations should be guided by the squared amplitudes of the branches. Since the squared amplitude of the branch in which you are alive 1,000 years from now is miniscule, you shouldn't expect to be alive 1,000 years from now.

However, David Lewis (2004, 17) suggests that there is something wrong with the account of rational expectation in this case. The basic point is that the branches in which you are dead shouldn't figure in your calculation of rational expectation, because there is nothing that it is like to be dead. In that case, the right way to figure what *you* should expect is first to ignore all the future branches in which you are dead, then renormalize the remainder (i.e., scale them so that their squared amplitudes sum to 1), and finally calculate the expectation value using these renormalized squared amplitudes. This automatically has the result that you should expect to be alive at any future time.

Incidentally, Lewis does not think that this is good news. As he notes, you avoid death, but not the normal effects of aging. Even though branches in which you avoid the effects of aging also exist at any future time, they form a miniscule proportion of the branches in which you are alive, and since you cannot discount branches in which you are alive, you should expect to age. You will live forever, but you will become increasingly decrepit: "You should expect to lose your loved ones, your eyes and limbs, your mental powers, and your health" (D. Lewis, 2004, 20). This is not a fate to be wished for.

But does many-worlds quantum mechanics really entail this fate? The motivation for Lewis's renormalization procedure is rather unclear. In the ordinary classical case, you don't discount future possibilities in which you are dead in calculating expectation values, even though you will not experience those possibilities. That is, if Lewis's renormalization argument works in the many-worlds case, it looks like it ought to work classically, too. And conversely, if it fails in the classical case, which it surely does, it must fail in many-worlds as well (Papineau, 2004).

Even so, it does feel like there is a difference between the many-worlds case and the classical case. This difference is perhaps best brought out by the following kind of thought experiment (D. Lewis 2004, 17). Suppose I have a Star Trek transporter that works by (painlessly) vaporizing you at one location, transmitting complete information about your physical make-up, and reconstructing you at another location. Suppose I configure this transporter so that it produces *two* copies of you at the other location rather than one. If you were unconcerned about your survival in the original transporter, clearly you should be unconcerned about your survival in the two-copy case. This is analogous to many-worlds branching. Now suppose that my two-copy transporter malfunctions, so that the second copy is never fully formed, but just appears as a cloud of dust. It still seems that you should be unconcerned about your survival; one copy was good enough for survival in the functioning transporter, so it is good enough now. This is analogous to many-worlds branching in which you are dead in one branch. And clearly the case of a malfunctioning two-copy transporter is very different from that of a malfunctioning one-copy transporter that produces a good copy of you half the time and a cloud of dust half the time; you should be very wary of stepping into this machine, because there is a 50% chance that you won't survive. This last case is analogous to classical risk-of-death situations.

Of course, the analogy between Star Trek transporters and many-worlds branching is not particularly close. In the Star Trek case, it is only *you* that ends up in two places, but in the many-worlds case the branches are whole worlds. This means that you cannot simply ignore the branches in which you are dead; they contain grieving successors of your relatives, for example. So risks of death are not to be taken lightly in branching worlds, and you should try to minimize such risks just as in classical worlds. But this is a separate issue from whether you should expect to *survive*, in the sense that your experience continues into the future.

Lewis's position on many-worlds immortality, perhaps not surprisingly, can be tied to his views on the identity of people. If people are identified with four-dimensional histories from birth to death, then the branches in which there is a corpse rather than a person don't figure into the total count of people before the branching event.[8] If probability can be understood as a measure of uncertainty about which premeasurement person I am (e.g., by applying the Ismael-Sider indexical reference strategy described earlier), then Lewis's policy of ignoring branches in which I am dead makes perfect sense.

But of course the most promising approach to probability in many-worlds does not take probability to be a measure of premeasurement uncertainty in this way. Indeed, as noted earlier, the Deutsch-Wallace approach to probability can be divorced from all talk of uncertainty.[9] The branch-counting approach to probability—of which the person-counting approach is a straightforward variant—violates Wallace's branching indifference axiom. Counting persons faces the same conceptual difficulties as counting branches, but if these difficulties can be overcome in the way explored in the previous section, then the viability of the person-counting approach *undermines* the Deutsch-Wallace argument, and hence undermines the viability of the many-worlds approach.

In other words, immortality should not be seen as a consequence of many-worlds quantum mechanics, because the strategies for making the immortality conclusion follow from the many-worlds ontology undermine the ability of the many-worlds theory to account for probability. Conversely, if the Deutsch-Wallace strategy for accommodating probability within the many-worlds theory succeeds, then you should not expect to live forever. The immortality argument is perhaps best viewed as a dramatic demonstration of the fundamental conflict between branch-counting (or person-counting) intuitions about probability and the decision-theoretic approach. The many-worlds theory, to the extent that it is viable, does not entail that you should expect to live forever.

6.4 Free Will

So what have we learned about determinism? Nothing conclusive: The quantum world might be straightforwardly deterministic (Bohm's theory), or straightforwardly indeterministic (GRW theory), or it might be that nonepistemic probabilities can emerge from a deterministic underlying physics (many-worlds theory). But at least we can say that the live options regarding determinism are much broader than they were according to classical mechanics. Does this broader range of options mean that we are in a better position to reconcile free will with quantum mechanics than with classical mechanics?

Historically, the greatest challenge to free will has been determinism. If the physical world is like a giant clockwork, what room is there for free human action? And as physics progressed, the more the universe seemed to resemble a giant clockwork. So it is not surprising, then, that the advent of quantum mechanics was hailed by many as giving a physical underpinning for free will, since quantum mechanics provides the first suggestion that determinism might fail at the fundamental physical level.

But quantum indeterminism is a double-edged sword. On the one hand, it loosens the grip of prior causes over future events. But on the other hand, the form that indeterminism takes in quantum mechanics looks pretty inhospitable to free will, too. The empirical adequacy of quantum mechanics requires that the outcomes of measurements be randomly distributed according to the Born rule, and randomness looks no more compatible with free will than determinism.

These points are commonplace and make it look like quantum mechanics has no special consequences for free will. However, some recent libertarian accounts of free will have embraced the randomness at the heart of (some versions of) quantum mechanics. According to Kane (1996), for example, a free action is one for which the agent is ultimately responsible, where ultimate responsibility requires that the character traits that led to the action were ultimately the result of actions that were not causally determined.

Such "self-forming actions" involve competing self-conceptions, and even though the choice between them is produced by an indeterministic (random) process, the agent is still responsible for the choice insofar as the choice is part of her striving toward the relevant self-conception.

I do not wish to debate the merits of such accounts here; my goal is only to note that there are accounts of free will that require indeterminism at the physical level. But as we have seen, it is unclear whether quantum mechanics can deliver the required indeterminism. The GRW theory is indeterministic in just the required sense: The outcomes of certain physical processes—those in which the position of a macroscopic object becomes correlated with the properties of a microscopic system—are generated by a fundamentally probabilistic dynamical law rather than purely by the prior state of the system. Bohm's theory does not contain this kind of indeterminism: The outcome of every physical process is fully determined by its prior state via deterministic dynamical laws. So in this sense libertarian accounts of free will like Kane's are empirically risky: They rely on an indeterministic interpretation of quantum mechanics winning out over its deterministic rivals.

As noted in Chapter 3, the GRW theory might be regarded as *particularly* empirically risky, since it postulates a physical collapse process for which we have no empirical evidence, and which may eventually be ruled out by careful interference experiments. It is perhaps unwise to bet on such a theory as the foundation for free will. It would be safer if both the GRW theory and the many-worlds theory could provide indeterminism in the required sense. But the many-worlds theory is deterministic at the level of physical law. Could it nevertheless play the same role as the GRW theory in underpinning an account of free will like Kane's?

I think the answer is "yes." Consider a case in which an agent has two competing self-conceptions, and two corresponding possible actions furthering one self-conception each. Under a GRW-type indeterministic story, the choice of action somehow becomes correlated with a microscopic system in the brain. If the microscopic system is in one state, one action is chosen, but if it is in another state, the other action is chosen. But in fact the microscopic system is initially in a superposition of the two states, so the Schrödinger dynamics results in a superposition of the two actions. Since these two actions presumably differ in the positions of some macroscopic objects, this superposition state is unstable, according to the GRW dynamics, and rapidly (and indeterministically) collapses to one choice or the other.

The many-worlds theory tells essentially the same story, except that the superposition of two distinct actions results in branching rather than collapse. Each action occurs in its own separate branch of reality. But each of the resultant people can ask themselves, "What is it that resulted in me making *this* choice rather than the other one," and in each case the answer has to be "Nothing." That is, for each postchoice person, there is nothing in the prechoice physical state that necessarily leads to one choice rather than the other.

The Lewisian view of people discussed earlier might lead to a slight worry on this score, though. Isn't it the case, on the Lewisian view, that there are two prechoice people, one of whom will make one choice and the other of whom will make the other? So isn't it already determined which choice each will make? Not really. Note that the individuation of people in this case is precisely on the basis of this future choice. So trivially, then, one person will make one choice and the other will make the other. But this no more determines what each person will do than the existence of a future fact about what a person will choose determines her choice in the GRW case. Of course, some people do get exercised by this kind of "logical determinism" and claim that it rules out free will (e.g., Taylor, 1962). But if your concern is with causal determinism, then it is no more present in many-worlds than in GRW.

So quantum mechanics *might* help the libertarian's case, depending on how things turn out. But most of us have compatilibilist inclinations.[10] Does quantum mechanics have anything to teach the compatibilist? One might think not, since compatibilism is consistent with both determinism and indeterminism, as long as agents have sufficient control over their choices. Certainly compatibilists don't have to side with many-worlds and GRW over Bohm; they can be equally happy with the latter. But there is a further possibility that is worth discussing here, as it suggests a sense in which quantum mechanics might be a threat even to compatibilism.

That further possibility, explored briefly in Chapter 5, is the program that seeks to evade the conclusion of Bell's theorem by violating his independence assumption—the assumption that says that the properties of a particle are independent of the measurements that will be performed on it. In this way, one might hope to construct a hidden variable theory without the nonlocal causation inherent in Bohm's theory. Bell calls such theories "super-deterministic" (Davies & Brown, 1986, 47), and reports that he is unable to take them seriously because of the damage they do to free will (Bell, 2004, 154).[11] The worry is that the violation of independence interferes with the experimenter's ability to freely choose which measurements to perform on a particle. This is easiest to see in the extreme case of a perfect correlation between the particle properties and the subsequent measurement. For example, suppose that, given the properties of a given particle, measurement X must be performed on it later. One might think that this interferes with the experimenter's ability to freely choose whether to perform experiment X or not. And even if the correlation is less than perfect, the correlation might be thought to interfere with the freedom of the experimenter to decide how frequently to perform measurement X on this kind of particle.

Prima facie, though, the correlations involved here seem no more problematic for compatibilism than those entailed by simple determinism. In a deterministic universe, the physical state of the world right now fixes which measurements I will choose to perform later, but this in no way restricts my freedom according to the compatibilist. So is Bell's worry misplaced? I think there is a more serious worry, and it has to do with the mechanism by which the correlation between the particle properties and the measurements performed on it is brought about. Consider first the situation in a deterministic universe. Although the state of the world right now fixes what measurements I will perform later, the way it does this is via *me*. That is, the deterministic causal chain passes through those mental and physical processes that constitute my free control over what I do; hence, there is (according to the compatibilist) no conflict between determinism and free will.

Now consider how the correlations involved in the violation of independence could be brought about. As discussed in Chapter 5, there are two options: A common cause in the past, requiring a vast hidden causal order, or backward-in-time causation from the later measuring events to the earlier particle production event. Let us take them in turn. Would the existence of a vast hidden causal order threaten compatibilist free will? It depends on the nature of the conspiracy. Perhaps the conspiracy is "shallow": It directly affects the immediate causal precursors of the measurement choice and the particle properties so that they match. In that case, my choice of measurement is to some extent out of my control, and the compatibilist strategy is weakened. But perhaps the conspiracy is "deep," in the sense that it affects the more distant causal precursors of my choice and of the particle properties. In that case the compatibilist strategy works as well as in the simple deterministic case: My choices are under my control, but surprisingly some of the factors involved in the formation of my volitions are correlated with the properties of particles I end up measuring.

But as mentioned in Chapter 5, appealing to a common cause mechanism is wildly implausible, and any threat to free will is the least of its problems. A more promising strategy is to postulate that the measurement performed on a particle affects its *earlier* properties—that is, to endorse backward-in-time causation. There is no need for a conspiracy in a retrocausal approach of this kind. Furthermore, there is no direct threat to compatibilist free will: My choice fixes the measurement performed in the usual forward-causal way, and that measurement influences the earlier properties of the particle I measure in a novel backward-causal way. So my choice is unconstrained.

Still, there is an indirect constraint on free will. Recall from Chapter 2 that Bell's theorem involves unexpectedly strong correlations between the spins of entangled particles, and the retrocausal theory explains these correlations by appealing to further correlations between the particles' properties and the measurements performed on them. Normally, you would expect to be able to destroy these latter correlations by freely choosing the measurement to be performed. But you can't. According to the retrocausal theory, any such attempt is doomed to fail. So there is something you might ordinarily take yourself to have control over that you don't have control over. This is a restriction on your (compatibilist) free will. But note that this restriction is limited to Bell-type experiments; you still have control over all your everyday choices, and overall the compatibilist approach to free will is unscathed.[12]

All in all, then, quantum mechanics is surprisingly uninformative about free will. It *might* help out a particular kind of libertarian, and it *might* entail a novel (but unimportant) constraint on compatibilist free will. But other than that it seems to leave the debate exactly as it was before.

6.5 Conclusion

Does God play dice? The evidence from quantum mechanics is equivocal. This is a clear case of underdetermination at work: Of the leading contenders, one (Bohm's theory) says the universe is deterministic, another (GRW) says it is fundamentally probabilistic, and in the case of the third (many worlds) it is hard to tell. This is not because the many-worlds theory is underspecified, but because it is genuinely hard to see whether there is a place for probability in a branching universe. Unless this issue can be resolved, the many-worlds theory is in trouble: If it can't be made to yield probabilistic predictions matching the Born rule, it can't reproduce the empirical success of standard quantum mechanics.

Part of the fascination with the issue of determinism in quantum mechanics can be explained by the hope that quantum indeterminism can somehow help reconcile free will with a scientific worldview. Much of this hope is probably misguided: Indeterminism is no magic bullet, since randomness is prima facie anathema to free will, too. But nevertheless, there *are* reasonable research programs on free will that depend on the existence of genuine indeterminism such as that exhibited by the GRW theory. And the "superdeterministic" approaches to quantum mechanics embodied by retrocausal approaches to quantum mechanics are *relevant* to compatibilist accounts of free will, although they arguably don't entail any serious threat to compatibilism.

7| Dimensions

UP TO NOW, THE METAPHYSICAL topics I have covered have been ones for which it is fairly commonplace to think that quantum mechanics has something distinctive to say. The topic of the current chapter, on the other hand, is one which has received surprisingly little attention until recently. The topic is how many spatial dimensions the world has according to quantum mechanics.

Here is why this is a significant issue. The three major approaches to quantum mechanics we have been discussing—Bohm's theory, the GRW theory, and the many-worlds theory—all (prima facie) interpret the wave function as describing a real physical entity.[1] According to the GRW theory and the many-worlds theory, this entity is all there is, and while in Bohm's theory there are also particles, the wave function (apparently) plays an essential dynamical role in pushing the particles around. The wave function describes a spread-out entity, but as noted in passing in Chapter 1, it is not spread out over ordinary three-dimensional space like the electromagnetic field. Instead, it is spread out over *configuration space*, a space that has three dimensions for each particle in the system we are modeling. So a one-particle system is modeled by a wave function that is distributed over the familiar three dimensions, a two particle system is modeled by a wave function spread over six dimensions, a three-particle system by a nine-dimensional wave function, and so on.

Furthermore, as far as ontology goes, the only system that really counts is the universe as a whole; every other choice involves an arbitrary distinction between system and environment. There are well over 10^{80} particles in the observable universe. So that means that the wave function of the universe lives in an extremely high-dimensional space. If the wave function describes a real physical entity, then the dimensionality of the physical world is much higher than we usually take it to be.

This is very interesting but also quite puzzling. If the physical world isn't really three-dimensional, why does it look like it has three spatial dimensions? I consider three possible lines of response. The first is that the three-dimensionality of the world is a kind of *illusion* created by the dynamical laws obeyed by the high-dimensional wave function. The second is that the world really is three-dimensional, and the wave function doesn't represent the spatial structure of the world as directly as we have been assuming. The third is that the world really is three-dimensional, and the wave function doesn't represent the physical world *at all*, but instead has some other job to do. But before we embark on these responses, let us remind ourselves why the wave function has to be represented in a high-dimensional space.

7.1 Configuration Space

Configuration space, as its name suggests, is a space of particle configurations. Each point in configuration space represents an arrangement of particles in ordinary three-dimensional space. So, for example, for a two-particle system, if particle 1 is at point $(1,3,4)$ in three-dimensional space, and particle 2 is at point $(9,2,5)$, then the arrangement of particles can be represented by the point $(1,3,4,9,2,5)$ in a six-dimensional configuration space.

The configuration space representation for particles is entirely a matter of convenience: An arrangement of two particles can be represented equally well by two points in a three-dimensional space or by one point in a six-dimensional space. But the same is not true of fields. Since a field ascribes numbers to each point in space, it is most naturally represented by a function of the three spatial dimensions. Suppose that the system we are modeling consists of two fields represented by the (real-valued) functions $f_1(x,y,z)$ and $f_2(x,y,z)$ defined over three-dimensional space. One possible configuration of the fields is illustrated in Figure 7.1, where the three spatial coordinates are represented schematically by the single horizontal axis. Here field 1 is nonzero only in regions A and B, and field 2 is nonzero only in regions C and D.

We could also represent this situation using a single function $F(x_1, y_1, z_1, x_2, y_2, z_2)$ defined over a six-dimensional space, where (x_1, y_1, z_1) are the coordinates over which function 1 is defined and (x_2, y_2, z_2) are

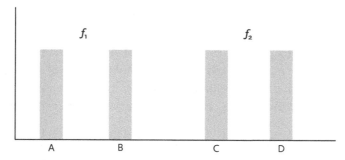

FIGURE 7.1 Field intensity in three-dimensional space.

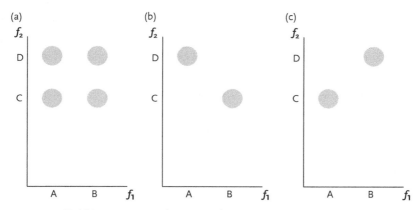

FIGURE 7.2 Field intensity in six-dimensional space.

the coordinates over which function 2 is defined. Such a representation is pictured in Figure 7.2 (a), where the three spatial coordinates of f_1 are represented schematically by the horizontal axis, and the three spatial coordinates of f_2 by the vertical axis. Again, the field is nonzero only in regions A and B in the coordinates of f_1 and in regions C and D in the coordinates of f_2. But note that exactly the same is true of the field distributions over configuration space shown in Figures 7.2 (b) and (c); each of them generates the two fields of Figure 7.1 when represented in three dimensions. That is, there are *several* configuration space representations corresponding to a *single* three-dimensional representation.

What's more, the configuration space representations contain information that is absent in the three-dimensional representation. In particular, the three-dimensional representation does not tell you about the correlations between the two fields. The field distribution shown in Figure 7.2 (c) is one in which field f_2 is large in region C if and only if field f_1 is large in region

A. The distribution shown in Figure 7.2 (b) is one in which field f_2 is large in region C if and only if field f_1 is large in region B. And the distribution shown in Figure 7.2 (a) is one in which there are no correlations between the amplitude of f_2 in regions C and D and the amplitude of f_1 in regions A and B. If this correlation information is significant, then the three-dimensional representation is inadequate, and the configuration space representation must be used.

This is exactly the situation in which we find ourselves in quantum mechanics. The state of a two-particle quantum system is represented by a (complex-valued) function over the six-dimensional configuration space—the wave function—and the correlational information implicit in the configuration-space representation is crucial to the empirical adequacy of the theory. If the wave function is as represented in Figure 7.2 (c), then the Born rule tells us that particle 2 will be found in region C if and only if particle 1 is found in region A. If it is as shown in Figure 7.2 (b), then particle 2 will be found in region C if and only if particle 1 is found in region B. And if the wave function is as shown in Figure 7.2 (a), then there is no correlation between the location of particle 2 and the location of particle 1. So the two-particle wave function in configuration space is not reducible to two one-particle wave functions in three-dimensional space; the configuration space representation of the wave function is apparently indispensable in quantum mechanics.

7.2 Three-Dimensionality as an Illusion

The fact that the wave function lives in a high-dimensional space was mentioned in passing by Schrödinger (1926b, 526), Bohm (1957, 117) and Bell (1981, 625; 1987b, 44), but it was introduced as a metaphysical puzzle by Albert (1996, 2013). Albert notes that according to the Bohm, GRW, and many-worlds theories, the wave function has the form of a field, and realism about a field commits you to realism about the space occupied by the field, since the field values are generally interpreted as intrinsic properties of the points of the space (1996, 278). For example, a realist attitude to electromagnetic theory commits you to the existence of a three-dimensional space whose points have the intrinsic properties ascribed to them by the electromagnetic field. So since the wave function is defined over configuration space, a realist attitude to quantum mechanics commits you to the existence of a $3N$-dimensional space, where N is the number of particles in the universe. Realism about quantum mechanics apparently leads to the conclusion that the physical world has many, many more spatial dimensions than we suspected.

While there is nothing intrinsically problematic about high-dimensional spaces and high-dimensional physical objects, it is immediately puzzling why the physical world appears to us to be three-dimensional if its dimensionality is much higher. Albert proposes that the answer is dynamical: The laws of our world are such that it will look to its inhabitants as if it has three spatial dimensions, even though it does not. Recall that the basic dynamical law of quantum mechanics is the Schrödinger equation:

$$i\hbar\frac{\partial}{\partial t}|\psi\rangle = \hat{H}|\psi\rangle. \tag{7.1}$$

All the realist accounts of quantum mechanics incorporate this law, although the GRW theory supplements it with a collapse dynamics, and Bohm's theory supplements it with a particle dynamics. In this equation, $\hat{H}$ is the operator for the total energy of a system, called the Hamiltonian. Classically, the total energy of a system depends on a number of factors, including the relative spatial positions of the particles, since the total energy of a system depends on the gravitational and electrostatic forces the particles exert on each other. But except in Bohm's theory, there are no particles at the basic level of description in quantum mechanics, and even in Bohm's theory, the motion of the wave function is independent of the locations of the particles. So in the quantum case, the total energy has nothing directly to do with the locations of any entities in three-dimensional space.

Nevertheless, Albert argues that the quantum Hamiltonian gives rise to the *appearance* of three-dimensionality. His point is that the Hamiltonian takes a particularly simple form if the $3N$ spatial coordinates are grouped into N sets of 3 (rather than $3N/2$ sets of 2, or $3N/4$ sets of 4, etc.). Even in quantum mechanics, there is a sense in which classical behavior emerges from quantum mechanical behavior in the macroscopic limit—that is, as systems become large and complicated. The sense is that while microscopic systems must typically be represented by a spread-out wave function in configuration space, macroscopic systems can always be represented to a good degree of approximation by a *point* in configuration space. And a point in $3N$-dimensional configuration space, as shown earlier, can equally well be represented as N points in three-dimensional space. Of course, it can also be represented as $3N/2$ points in a two-dimensional space, and so on. But if we choose to group the coordinates into threes, then the Hamiltonian takes a particularly neat form: The potential energy term depends only on the distances between the N "particles." On the other hand, if we choose to group the coordinates into twos or fours or sevens, then the potential energy term will bear no straightforward relation to the "interparticle" distances so produced.

The key point is that the form of the Hamiltonian one obtains by grouping into threes corresponds to our classical worldview according to which potential energy is generated by forces that depend only on the three-dimensional distances between the objects involved. If Albert is right, we have to give up the classical three-dimensional picture, since there are no particles moving in three-dimensional space and hence no laws governing their motion. But his claim is that we can nevertheless explain why our intuitive picture of the world has three dimensions, since it is the obvious (though false) interpretation of the behavior of macroscopic objects. That is, even though there is nothing in fundamental reality corresponding to our choice of coordinate grouping, if we choose to group the coordinates in threes, a particular description of the behavior of medium-sized everyday objects becomes available to us—namely the classical description of objects moving in three-dimensional space subject to forces that depend on the distances between them. Hence, "quantum-mechanical worlds are going to appear (falsely!) to their inhabitants, if they don't look too closely, to have the same number of *spatial dimensions* as their *classical counterparts* do," namely three (Albert, 1996, 282). The reason for the caveat, of course, is that if we *do* look closely—if we perform experiments that reveal the underlying quantum-mechanical nature of microscopic reality—then we convince ourselves that the world can't really be three-dimensional, via the arguments of the previous section.

Albert's argument applies to all three of the major interpretations of quantum mechanics. The argument applies most directly to the many-worlds theory, according to which there is nothing but the wave function, and the dynamical law is precisely the Schrödinger equation. According to the GRW theory, although there is nothing but the wave function, evolution according to the Schrödinger equation is interrupted occasionally by the collapse dynamics. Nevertheless, it is still true that there are no three-dimensional objects at the fundamental level of description in the GRW theory. For emergent macroscopic objects, while the collapse dynamics plays just as important a role in the evolution of the state over time as the Schrödinger dynamics, the former contains no mention of the energy properties of the system. So insofar as the macroscopic world approximates objects obeying the familiar laws of motion, that fact is accounted for by the special form of the Hamiltonian in the Schrödinger equation when the coordinates are grouped into threes, just as Albert argues.

The case is somewhat trickier with regard to Bohm's theory, since Bohm postulates a set of particles in addition to the wave function. However, since the wave function lives in configuration space, and the Bohmian particles are pushed around by the wave function, presumably the set of particles is a configuration space object, too—represented by a single point in configuration space. So the ontology of Bohm's theory resides in a high-dimensional space. Furthermore, the Bohmian law according to which this single high-dimensional "particle" moves makes no mention of the energy properties of the system; again, the explanation of the motion of macroscopic objects according to the familiar laws of motion is accounted for by the special form of the Hamiltonian in the Schrödinger equation when the coordinates are grouped into threes.

So Albert concludes that however you interpret quantum mechanics, "the space we live in ... is configuration space. And whatever impression we have to the contrary ... is somehow flatly illusory" (1996, 277). This is a radical conclusion. Other theories have postulated more than three spatial dimensions: for example, string theories postulate nine or ten spatial dimensions. However, since all but three spatial dimensions are curled up so small as to be insignificant at all but the smallest scales, there is a straightforward sense in which the world of string theory is three-dimensional: The familiar three spatial dimensions are intrinsically distinct from the rest and are the only three that show up at macroscopic scales. Quantum mechanics, on the other hand, doesn't pick out three dimensions as distinct from the rest. Furthermore, the dimensionality of the world is much less obviously a matter of debate and discovery than, say, determinism; if quantum mechanics shows that the world is not really spatially three-dimensional, this is much more surprising than any failure of determinism would be.

7.3 Adding Ontology

Maybe Albert's conclusion is not just radical, but untenable. That is, you might think that the underlying ontology of quantum mechanics (or any other theory) cannot be other than three dimensional.[2] You might hold this view for one of several reasons. First, you might think that Albert's argument that a high-dimensional universe can look to its inhabitants as if it is three-dimensional fails (Monton, 2002). Note that there is no objectively preferred grouping of the coordinates into threes; the fact that the Hamiltonian takes a simple form under a particular choice is a pragmatic matter concerning which grouping is easiest to use. There is no reason to think that the way the world looks when we group the coordinates into threes is any more a reflection of reality than the way it looks when we group the coordinates into sevens. But the familiar objects of experience—tables and chairs, buildings and people—only appear under the three-dimensional grouping. So it looks like we are forced to say that there aren't really any tables or chairs or buildings or people—that they are artifacts of our choice of coordinate grouping. Now, though, the explanation of our experience of the world as three-dimensional starts to looks suspect: What exactly is it that we are explaining? You can't say that we are explaining the fact that tables and chairs and people appear to be three-dimensional objects if it is not an objective fact that there are tables and chairs and people. If it is not an objective fact that there are people, it is not even clear that it makes sense to say that *we* experience the world as three-dimensional.

I don't think such arguments succeed (P. Lewis, 2004). For one thing, it is possible to bite the bullet here: Even if (objectively speaking) there are no tables or chairs or people, it is still arguably enough for a theory to generate appearances as if there are such three-dimensional objects. We are not (genuinely) three-dimensional, and nor are tables and chairs, but it will look to us as if we are three-dimensional objects in a three-dimensional world. (And I will argue in the next section that you don't really have to bite this bullet.) But even if you are not convinced by this particular line of reasoning, you might think that there are general reasons why the entities most directly underlying our experience should be genuinely three-dimensional.

For example, Maudlin (2013) argues from an epistemic perspective: What we directly observe are local matters of fact, where "local" is to be read as "in a restricted region of three-dimensional space." The job of a scientific theory is to explain these observations, and the only possible explanation is in terms of other local matters of fact. That is, at least that part of the theory that directly accounts for our observations must be written in terms of the behavior of genuinely three-dimensional entities. Maudlin appeals to examples from the history of science to support his argument. Allori (2013) makes a more direct appeal to the historical development of scientific theories: As science progresses, we make incremental, minimal modifications to our conception of the underlying ontology. This can be justified in terms of methodological conservatism, and in terms of a norm that says that our scientific ontology should stay as close to our everyday ontology of macroscopic objects as the empirical phenomena permit. On this view, Albert's leap beyond a three-dimensional ontology is unwarranted as long as three-dimensional alternatives are available.

And both Allori and Maudlin think that three-dimensional alternatives *are* available. The basic idea is to reject Albert's implicit assumption that the appearances are to be explained in terms of the wave function alone; instead, we should supplement the wave function with additional *primitive ontology* that is genuinely three-dimensional, and in terms of which our observations can be explained.[3] The most obvious candidate for an ontological addition of this kind is the set of particles posited by Bohm's theory. Note that this is not Albert's version of Bohm's theory, in which the ontology consists of a field plus a single point inhabiting the same $3N$-dimensional space, but a version in which there is a field in configuration space plus N points moving in a genuinely three-dimensional space. Explaining our experience is straightforward on this model: A three-dimensional macroscopic object is just an arrangement of Bohmian particles in three-dimensional space.

As Allori and Maudlin both note, this is not the only way to generate a genuinely three-dimensional ontology. One could instead begin with the GRW theory and supplement the field described by the wave function with a distribution of mass density over three-dimensional space, where the mass density in a small three-dimensional region corresponds to the expected mass density in that region predicted by standard quantum mechanics.[4] In this case, a macroscopic object is just a region of relatively high mass density in three-dimensional space. Alternatively, one could supplement GRW with a set of *flashes*, where a flash is an instantaneous event in three-dimensional space at the center-point of a GRW collapse event. Here a macroscopic object is constituted by an arrangement of flashes in three-dimensional space over a particular period of time (Tumulka, 2006a). These are the "massy" and "flashy" versions of GRW mentioned in passing in Chapters 4 and 5.

A final possibility is to supplement the wave function of standard quantum mechanics (i.e., without a collapse process) with a mass distribution over three-dimensional space defined as in massy GRW; in this way one obtains a massy version of the many-worlds theory (Allori, Goldstein, Tumulka, & Zanghì, 2011). Here one has a branching structure of mutually noninteracting three-dimensional mass density distributions, and a macroscopic object is a region of relatively high mass density within one such branch. The key feature of each of these theories is the specification of a genuinely three-dimensional ontology underlying the appearances. So there is no need for a dynamical explanation of the apparent three-dimensionality of the world; the world appears three-dimensional because (at least this aspect of) it *is* three-dimensional.

However, there are a number of problems with this approach (North, 2013, 192). The first is that it appears somewhat ad hoc: The only reason to add this additional ontology to standard quantum mechanics is to avoid Albert's radical conclusion. This appearance can be lessened somewhat by changing the order of exposition: Rather than starting with standard quantum mechanics and adding extra ontology, start with the theory that posits three-dimensional ontology. One way of expressing the trouble with quantum mechanics is that it was formulated without any ontological picture in mind (Chapter 1). If it is granted (as Allori and Maudlin insist) that we should always begin the process of theory construction by positing some three-dimensional ontology that the theory is about, then each of the earlier proposals looks reasonable.

But it is unclear whether we should grant their assumption about the need for three-dimensional ontology. Construed as an a priori principle constraining theory construction, it seems too strong: The scientific image has departed from the manifest image in a number of significant ways, so why insist that we may not depart from the manifest image when it comes to spatial dimensions? Construed as a pragmatic principle of methodological conservatism, it seems like it might be too weak: Quantum mechanics without the additional ontology is much simpler as a theory, so maybe the time has come to give up our insistence on fundamental three-dimensionality.

Furthermore, there is a general structural problem with all the three-dimensional additional ontologies just described: The additional ontology is dependent on the wave function, but the wave function is defined over a high-dimensional configuration space and the additional ontology is defined over three-dimensional space. How are the two spaces related to each other? This is precisely why Albert assumes that the additional ontology of Bohm's theory is a single "particle" residing in the same high-dimensional space as the wave function. If not, and the wave function and the particles reside in different spaces, then how can the former direct the motion of the latter? How can the mass density or flash distribution in three-dimensional space be determined by a field in configuration space? We need such an account if the addition of three-dimensional ontology is to do the job assigned to it.

7.4 Interpreting the Wave Function

Allori and Maudlin are both fully aware of this problem. Their preferred solution in the case of Bohm's theory is to argue that while the wave function, qua mathematical function, inhabits a high-dimensional space, we do not have to interpret the mathematics as representing a real high-dimensional field. As Maudlin (2013, 145) points out, we should be "tentative and cautious" about the interpretation of the wave function, precisely because (unlike the three-dimensional primitive ontology), it is not directly responsible for the production of observable results. Since whatever it is that the wave function represents is only indirectly responsible for our observations, Maudlin recommends that we regard the wave function merely as fulfilling a particular functional role.

The functional role that both Maudlin and Allori suggest in the case of Bohm's theory, following Dürr, Goldstein, and Zanghì (1997), is that of a *law*. The wave function in Bohm's theory, according to this proposal, represents the law determining the motion of the Bohmian particles. The immediate problem with this proposal is that the wave function changes over time, whereas laws (presumably) don't change. But Dürr, Goldstein, and Zanghì note that certain attempts to reconcile quantum mechanics with general relativity lead to a *static* wave function for the universe, in which case it becomes more plausible to regard the wave function as representing a law.[5]

As Goldstein and Zanghì (2013) point out, it is still possible for the wave function representing a subsystem of the universe to change over time even if the wave function of the whole universe is static. For example, consider the wave function for a single particle; one obtains it from the wave function of the universe by substituting the actual positions of all the other particles for the spatial variables in the wave function of the universe. For example, if the wave function of the universe is represented by $\psi(x_1, y_1, z_1, x_2, y_2, z_2 \ldots)$, and the positions of the Bohmian particles are given by (X_1, Y_1, Z_1), (X_2, Y_2, Z_2), ..., then the wave function of particle 1 alone is represented by $\psi(x_1, y_1, z_1, X_2, Y_2, Z_2, X_3, Y_3, Z_3 \ldots)$. Even if the wave function of the universe has no time dependence, the wave function of particle 1 inherits a time dependence from the fact that the Bohmian particles move over time. This is promising, but it is worth noting that proposals for reconciling quantum mechanics with general relativity remain conjectural; there is no consensus on how to do this. So it is far from certain that the wave function of the universe will turn out to be static. If the wave function of the universe is time dependent, presumably it can't be interpreted as a law.[6]

Another functional role one might consider for the wave function is as epistemic—as merely a convenient summary of our knowledge of the properties of something that resides in three-dimensional space. Of course, laws have epistemic features, too, but the proposal here is different: Perhaps the three-dimensional ontology moves according to laws that make no reference to the wave function, so the wave function is simply a useful bookkeeping device expressing our best information about where the underlying entities are. This interpretation of the wave function is not available to Bohmians, since the motion of Bohmian particles depends on the wave function (and similarly for the massy and flashy ontologies). The no-go theorems discussed in Chapter 2 place stringent constraints on any purely epistemic interpretation of the wave function. But if one is prepared to admit backward causation, then as discussed in Chapter 5 it may be possible to construct a hidden variable theory in which there are only particles and their properties (e.g., Price, 2012), or only a classical wave inhabiting a three-dimensional space (Wharton, 2010). Then there would be no difficulty in interpreting the wave function: The high-dimensional representation is just a reflection of the conditional nature of a good deal of our knowledge of the underlying ontology. In the particle case, the wave function in Figure 7.2 (c) represents a state of knowledge in which I don't know whether particle 1 is at A or at B, and I don't know whether particle 2 is at C or at D, but I know that particle 1 is at A if and only if particle 2 is at C. However, retrocausal models of quantum mechanics are somewhat conjectural at this stage, too.

Even if the wave function is interpreted realistically—as representing something physical—one might wonder whether the high-dimensional structure of the wave function needs to be interpreted as a genuinely *spatial* structure in the world. Indeed, various realist interpretations of the wave function that deny a literal spatial reading of its structure have been suggested. Monton (2006, 2013), for example, suggests that the wave function should be regarded as representing a property of the set of particles in the universe, where those particles live in a three-dimensional space. Wallace and Timpson (2010) suggest that the quantum state should be regarded as representing a set of properties of three-dimensional spatial (or four-dimensional space-time) regions. In each case, the properties are unfamiliar and perhaps rather mathematically complicated, but as Wallace puts it, there is "no rule of segregation which states that, for example, only those mathematical items to which one is introduced sufficiently early on in the schoolroom get to count as possible representatives of physical quantities" (2012, 299). It may be convenient in a number of ways to write the wave function as a function over configuration space, but it could equally well represent complicated properties inhabiting three-dimensional space. Given Maudlin's admonition of interpretive caution, there is no reason to take the configuration space representation as a direct picture of the structure of the world (Lewis 2013a).

If this interpretive strategy is successful, then there is no need to regard the world as "really" high-dimensional, and hence no need for a special story about why the world looks three-dimensional to us: It looks three-dimensional because it *is* three-dimensional.[7] But by the same token, there is no need to supplement the quantum state with additional ontology in three-dimensional space, since the quantum state already represents three-dimensional ontology. Wallace (2012, 299) endorses a many-worlds approach: The branching state is instantiated in four-dimensional space-time. Monton's proposal is most naturally interpreted as a Bohmian one: There is an arrangement of particles in three-dimensional space, and the quantum state represents a property of this set of particles. But the general strategy can be extended to GRW-type theories, too: again, the quantum state at a time is instantiated by properties of three-dimensional spatial regions, and the collapse process means that unique well-defined locations for macroscopic objects can be constructed out of these properties (Lewis, 2006b).

Now we seem to have come full circle. Albert argues that the world of quantum mechanics is a high-dimensional world. Allori, Maudlin, and others respond that this is unacceptable—that an adequate physical theory must always start from a three-dimensional ontology. So they consider ways of adding three-dimensional ontology to the high-dimensional ontology described by the wave function. But if the wave function and the ontology underlying our experience live in different spaces, it becomes mysterious how they relate to each other. There are various ways to argue that in fact the spatial structure of the wave function is three-dimensional, so that in principle we could tell a story about how the wave function relates to the additional ontology. But why tell this story? The additional ontology—the mass density or the flashes—has become otiose, since the wave function doesn't describe a high dimensional reality after all.

7.5 Conclusion

Prima facie, the high dimensionality of the quantum wave function is a problem. If we insist that the wave function is a direct representation of the spatial structure of the world, then we are forced to accept that the world has many, many more spatial dimensions than we suspected, and we have to take on the burden of explaining our appearance of living in a three-dimensional world. Despite the arguments considered here to the contrary, I think that Albert's dynamical argument essentially succeeds at delivering such an explanation. But is such an explanation really necessary?

I think not. There seems little motivation for a direct spatial reading of the wave function. One can be a realist about the structure implicit in the wave function without thinking that every parameter is an independent spatial parameter. If we adopt a more flexible approach to interpreting the wave function, then it becomes clear that it can describe a three-dimensional world after all.

Nevertheless, the fact that the wave function is most easily represented in configuration space should not be overlooked. Processes that are entirely local in configuration space can be nonlocal in three-dimensional space, and properties of a single point in configuration space can be properties of spatially scattered systems in three-dimensional space. That is, the fact that the wave function is most easily represented in configuration space may tell us that quantum mechanics is fundamentally holistic or involves irreducible emergent properties. These matters will be explored in the following chapter.

8| Parts and Wholes

QUANTUM MECHANICS IS OFTEN SAID to entail a kind of holism—that wholes are in some sense more than just the sum of their parts. But this can appear to be a rather obscure position: What more could there be to a whole than its parts? Consequently, many are attracted to the crystalline clarity of some species of part-whole reductionism. Perhaps the most prominent example of reductionism is David Lewis's *Humean supervenience*, which he describes thus:

> It is the doctrine that all there is to the world is a vast mosaic of local matters of fact, just one little thing and then another.... We have geometry: A system of external relations of spatio-temporal distance between points. Maybe points of space-time itself, maybe point-sized bits of matter or aether fields, maybe both. And at those points we have local qualities: perfectly natural intrinsic properties which need nothing bigger than a point at which to be instantiated. For short: we have an arrangement of qualities. And that is all. All else supervenes on that. (Lewis 1986b, x)

That is, once you have specified the properties of the basic parts of a system—which Lewis takes to be point-like—you have fixed all the properties of the system, at every level of description.

Lewis describes Humean supervenience as "something I would very much like to believe if at all possible" (1986b, 111). It is attractive for its clarity, if for no other reason. The question before us is whether quantum mechanics rules out Humean supervenience—whether the quantum world can be understood in terms of an arrangement of local qualities or not.

In general, there are two ways in which the reduction of wholes to their parts might fail. First, there might be genuinely emergent properties—properties of the whole system that cannot be reduced to the properties of its parts. Second, and more radically, there might be no parts at all. I will take these possibilities in turn.

8.1 The Case for Holism

Consider an entangled state, for example a pair of particles in the state $|S\rangle = (|\uparrow\rangle_1|\downarrow\rangle_2 - |\downarrow\rangle_1|\uparrow\rangle_2)/\sqrt{2}$, where the spins are defined relative to the z-axis. This state can be regarded as describing a property of the two-particle system. The question is whether this property can be reduced to a property of particle 1 and a property of particle 2. So what are the properties of the individual particles in this situation? A flat-footed way to extract a property for particle 1 is simply to erase anything with the subscript "2," to obtain $|S_1\rangle = (|\uparrow\rangle_1 - |\downarrow\rangle_1)/\sqrt{2}$. Similarly, we obtain $|S_2\rangle = (|\downarrow\rangle_2 - |\uparrow\rangle_2)/\sqrt{2}$ as a putative property of particle 2.

But as noted in Chapter 7, the single-particle properties $|S_1\rangle$ and $|S_2\rangle$ do not entail state $|S\rangle$. In particular, the two-particle state $|S'\rangle = (|\uparrow\rangle_1 - |\downarrow\rangle_1)(|\downarrow\rangle_2 - |\uparrow\rangle_2)/2$ also reduces via this flat-footed technique to $|S_1\rangle$ and $|S_2\rangle$, even though $|S\rangle$ and $|S'\rangle$ are different states. The difference between them doesn't show up in the spins of the individual particles: If we measure the spin of particle 1, we obtain spin-up half the time and spin down half the time whether the state of the system is $|S\rangle$ or $|S'\rangle$, and similarly for particle 2. This is the sense in which $|S_1\rangle$ and $|S_2\rangle$ describe the individual particles correctly. But if the state of the whole system is $|S\rangle$, then the spin of particle 1 is always the opposite of the spin of particle 2, whereas if the state is $|S'\rangle$ there is no correlation between the spins of the two particles. $|S_1\rangle$ and $|S_2\rangle$ leave out this information. That is, the two single-particle states do not distinguish between two-particle states in which the spins are correlated and two-particle states in which they are not.

On this basis, Teller (1986) concludes that quantum mechanics exhibits *relational holism*, which he defines as the position that two individuals can have relational properties that do not supervene on their nonrelational properties. The idea is that the two particles share a relational property—that of having perfectly anticorrelated spins—that does not supervene on the non-relational spin properties of the individual particles. If we define an *emergent* property of a system as one that does not supervene on the properties of the individuals that make up the system, then relational holism amounts to the existence of emergent properties.

Teller notes that "holism has always seemed incoherent, for it seems to say that two distinct things can somehow be entangled or intermeshed so that they are not two distinct things after all" (1986, 73). Similar worries could be expressed about emergence. But the relational holism or emergence exhibited by an entangled state is perfectly clear and well stated. So quantum mechanics at least gives us a clear case that we can point to as an example of what kind of system might exhibit holism and emergence. That is, Teller takes it that quantum mechanics can turn holism from an obscure to a reputable philosophical notion.

However, as yet we have not looked very hard to find a way of reducing the entangled state to properties of the individual particles; we just examined the simplest flat-footed reduction and showed that it doesn't work. If we are to show that quantum mechanics *entails* holism, we will need to establish conclusively that there is no way of reducing the relational property of the two-particle system to the nonrelational properties of its parts. Hawthorne and Silberstein (1995) attempt to do just that, using Bell's theorem to establish the holistic conclusion. Recall from Chapter 2 that Bell's theorem establishes, subject to certain assumptions, that the empirical correlations we observe for entangled states cannot be reproduced by any recipe for ascribing spin properties to the individual particles. This looks like exactly what we are looking for: If the assumptions of Bell's theorem are met, then the explanation of the empirical results we see cannot be purely in terms of the individual properties of the particles. So if we want the results to be explainable at all, then we have to postulate an emergent property of the two-particle pair—a property that is not reducible to the properties of the individual particles.

But of course the assumptions on which Bell's theorem is based do not go without saying, and Bell himself took the lesson of his theorem to be that at least one of them must be false. The assumptions in question are independence—that the properties of the individual particles are independent of the measurements performed on them later—and locality—that the performance of a measurement on one particle does not affect the properties of the other. Hawthorne and Silberstein argue that any mechanism that circumvents Bell's conclusion by violating independence must either violate locality or involve emergent properties (1995, 128). Further, they contend that locality cannot be violated, since the influence of the measurement of one particle on the properties of the other would have to travel faster than light, and faster-than-light influences are forbidden by special relativity (1995, 132). Hence, they conclude that quantum mechanics, however it is interpreted, must involve emergent properties—that the phenomena of entanglement provide direct evidence of holism in the world.

However, this conclusion needs to be treated with some caution. First, there is a way of evading the conclusion of Bell's theorem without violating either independence or (arguably) locality, namely the many-worlds theory. Second, Hawthorne and Silberstein's argument concerning the violation of independence assumes that independence is violated in a particular way—by a common cause in the past of both the measurement event and the particle production event. Further, they assume that such a cause cannot itself depend on the types of measurement performed on the particles. But we saw in Chapter 5 that the postulation of a common cause is not the most plausible way to violate independence. A better way to go is to propose that the measurement event itself can affect the earlier properties of the particle via backward causation. But in that case Hawthorne and Silberstein's assumption that the cause cannot depend on the type of measurement performed is no longer secure, as the event providing the causal link between the choice of measurement and the particle properties is precisely the measurement itself.[1] Given the possibility of a retrocausal violation of independence, Hawthorne and Silberstein's argument that a violation of independence must also involve either nonlocality or holism cannot be relied on.[2] Finally, Hawthorne and Silberstein's injunction against violating locality may not be secure either. We saw that Bohm's theory and the GRW theory violate locality, and at least some authors have suggested that we can accommodate this by adding an absolute standard of simultaneity to the space-time structure of special relativity (Chapter 5).

So it looks like the prospects for establishing holism on the basis of Bell's theorem are rather dim. Perhaps there is no general argument from the quantum phenomena to a holistic conclusion. However, we can still make some progress by following the procedure adopted elsewhere in this book—namely by considering the various distinct interpretations of quantum mechanics individually. To what extent do each of the major interpretations of quantum mechanics incorporate a holistic ontology? This is the business of the following section.

8.2 Holism Reconsidered

The three major interpretations of quantum mechanics violate the assumptions of Bell's theorem in different ways. Let us start by considering the theories that violate locality, namely Bohm and GRW. Does the existence of causal influences between the measurement on one particle and the properties of the other obviate the need for holism? You might think that it should. If particle 1 can (in principle) send any kind of instantaneous influence to particle 2, then when particle 1 is measured in the z direction and found to be z-spin up, it can send a signal to particle 2 saying "I am z-spin up; adjust your properties accordingly." Particle 2 can respond by adopting a 100% disposition to be z-spin down on measurement, and a 25% disposition to be spin-down in the other two directions involved in the Bell experiment.

But in fact this doesn't work. The reason it doesn't work is that the signal sent by particle 1 has to be directed solely at particle 2. Suppose we prepare two pairs of entangled particles, where particle 1 is entangled with particle 2 as before, and particle 3 is entangled with particle 4. If particle 1 is measured and found to be z-spin up, particle 2 must be z-spin down, but there is no correlation between the spin of particle 1 and the spins of particles 3 and 4. However, if particle 1 broadcasts a causal influence on measurement, it will equally be received by particles 2, 3, and 4. If only particle 2 is to be receptive to this signal (and only particle 4 is to be receptive to a signal sent by particle 3), then particles 1 and 2 (and particles 3 and 4) must somehow be linked prior to the measurement. This is the role of the relational property of the pair.

You might still try to construct this link using nonrelational properties of the individual particles, and in a sense it is possible. You can ascribe to each of particles 1 and 2 a nonrelational property that is unique to them—that is not shared with any other particles—and then stipulate that a signal sent by a particle with this property is only receivable by another particle with this property. But this is rather baroque: There will need to be a different property for every pair of entangled particles in the universe.[3] And it is really cheating, since this scheme is really just a way of encoding a relation. Consider, for example, a relation like "taller than." You could attempt to mimic the relational claim "Alice is taller than Bob" using only nonrelational properties by ascribing each of Alice and Bob a nonrelational property p that is not shared with any other people and also ascribing Alice the property "tall$_p$," and Bob the property "short$_p$."[4] But this is clearly a trick: "Taller than" is still a relation that holds between Alice and Bob. Similarly in the quantum case: Entanglement is still a relation that holds between a particular pair of particles.

So the fact that entanglement holds between a specific pair of particles means that nonlocal theories still need emergent properties. Bohm's theory and the GRW theory are nonlocal, but nonlocality cannot recover Humean supervenience. In Bohm's theory, the entangled state $|S\rangle$ cannot be fully captured by an ascription of nonrelational properties to the two particles involved, or to the nonrelational properties of the two regions of space occupied by their respective wave packets. In the GRW theory, the entangled state $|S\rangle$ cannot be fully captured by an ascription of nonrelational properties to the two regions of space occupied by the "particle 1" and "particle 2" wave packets. In each case, specifying the properties of the parts doesn't suffice to specify all the properties of the system; we also need to specify irreducible properties of the whole.

What about the many-worlds theory? Recall that the many-worlds theory evades the force of Bell's theorem by denying the tacit assumption that measurements have unique outcomes, rather than the explicit assumptions of locality and independence. In the case of a pair of particles in state $|S\rangle$, a measurement of the spin of particle 1 yields *both* outcomes, spin-up and spin-down, and similarly for particle 2. You might think that this would eliminate the need for holism, as each particle can simply be ascribed both spin properties. But this overlooks the role of many-worlds branching in producing the correlations we observe for entangled states. A z-spin measurement on particle 1 produces two branches, one containing an observer seeing spin-up and one containing an observer seeing spin-down. But a subsequent z-spin measurement on particle 2 produces no further branching: The branch in which the observer of particle 1 sees spin-up is also a branch in which the observer of particle 2 sees spin-down, and the branch in which the observer of particle 1 sees spin-down is also a branch in which the observer of particle 2 sees spin-up. So when the observers get together to compare results in their respective branches, they always find that they disagree.

We saw earlier that the correlations between the spin results for a pair of particles in an entangled state can only be explained via an irreducible property of the pair. Although the many-worlds theory denies that there are unique spin results, it does not deny that there are correlations between the branch-relative spin results for the two particles. Since it is the correlations that require the irreducible properties, not the unique outcomes, the many-worlds theory too requires irreducible properties of the pair. More precisely, no ascription of nonrelational properties to the two regions of space occupied by the "particle 1" and "particle 2" wave packets suffices to yield the correlations between the spin results. So just like GRW and Bohm, the many-worlds theory requires emergent properties and violates Humean supervenience.[5]

So all three of the major interpretations of quantum mechanics embody a form of holism, in violation of Humean supervenience. Can we regard holism as a consequence of quantum mechanics, then? Perhaps not. There may be other possibilities. First, note that up to now I have been assuming that Humean supervenience is to be formulated in three-dimensional terms; the local qualities to which it refers are those that can be ascribed to a point (or a small region) of ordinary three-dimensional space. But as considered at length in Chapter 7, the wave function of quantum mechanics is defined over a configuration space with three dimensions for each particle in the system, not over three-dimensional space. We saw in that chapter that we do not have to interpret this structure as telling us that the world really does have $3N$ dimensions (where N is the number of particles in the universe). But *if we do*, then a new way of defending Humean supervenience becomes available—and indeed, one might take this as a reason to accept the high-dimensional ontology (Loewer, 1996).

The idea is that in the $3N$-dimensional space, the reduction of the properties of wholes to the properties of their parts is entirely unproblematic, since correlations between distant locations in three-dimensional space can be represented as distributions of local properties over $3N$-dimensional space. For example, the wave function distribution over six-dimensional space shown in Figure 7.2 (b) in Chapter 7 is one in which the positions of two particles are correlated: It could be the state of a pair of particles in spin state $|S\rangle$ after the positions of the particles have been correlated with their spins, for example. The wave function distribution here is entirely reducible to the wave properties (amplitude and phase) ascribed to points of six-dimensional space, as the diagram directly demonstrates.

However, there are reasons to think that this strategy for rescuing Humean supervenience is less than satisfactory. The idea that the world is really spatially $3N$-dimensional is a radical one, and as we saw in Chapter 7, it is not otherwise forced on us. One may find it a high price to pay to save our intuitions concerning the reduction of wholes to parts. Furthermore, it is not clear that the strategy is entirely successful. Arguably, part of what we want of Humean supervenience is that the ascription of properties to one part of space is independent of the ascription of properties to another part of space. That is, the fact that a certain property is instantiated *here* should tell you nothing about the properties of other locations in space. But if that is part of Humean supervenience, then quantum mechanics violates Humean supervenience even when formulated in $3N$-dimensional space, due to the fact that the squared wave function amplitude must integrate to 1 over the whole of the space (Myrvold, 2015). In Figure 7.2 (b), for example, half the squared wave function amplitude is concentrated in the top-left corner, which tells you that it cannot be the case that more than half of the squared wave function amplitude is concentrated in any other region of the space. Hence, the properties of one region constrain the properties of a distant region, even in $3N$-dimensional space.

A second way one might try to rescue Humean supervenience in the context of quantum mechanics is by adopting the retrocausal approach described in Chapter 5. Recall that this approach satisfies Bell's locality assumption but violates his independence assumption: The properties of a particle can depend on the measurements performed on it. The way this happens is that, just as in ordinary causal theories particles can bear the traces of their past physical interactions, so in retrocausal theories particles can bear the traces of their *future* physical interactions, including measurement interactions.[6] So for a pair of entangled particles, a z-spin measurement on particle 1 affects the earlier properties of particle 1, and since the two particles have a common source, this can affect the properties of particle 2 as well; the particles are emitted at the source with opposite z-spins. A different spin measurement on particle 1 would result in the particles being emitted at the source with different spin properties.

So retrocausal quantum mechanics, unlike Bohm and GRW, has no need for irreducible relational properties of the pair of particles to explain the correlated measurement results. Humean supervenience is recovered; the explanation of the measured spin results is entirely in terms of the possessed spin property of each individual particle. And there is no worry about the ascription of properties to one location ruling out certain properties of another location in this case. *Causal laws* may make certain ascriptions of properties to space-time locations incompatible with each other, but this is true in the classical case too: You can't ascribe a classical object the property of being here one second and being on the moon the next. But this is generally seen as no threat to Humean supervenience, since Humeans typically regard property ascriptions as logically prior to the laws: First arrange the local qualities over space and time, then read the laws off the result. If we can set aside the laws when considering which arrangements of properties are allowed, then the properties ascribed to one location don't constrain those ascribed to another location.[7]

If this retrocausal strategy is successful, then the entangled quantum state $|S\rangle$ can be regarded as a reflection of the *epistemic* state of an agent who doesn't know what measurements will be performed on the two-particle system. What the agent knows can only be expressed holistically of the pair of particles: The various correlations between the results of possible spin measurements cannot be expressed in terms of properties of the individual particles, as we have seen. But nothing follows from this epistemic holism about the actual properties of the particles. Again, though, backward causation might be thought to be too high a price to pay to save part-whole reduction, and we have no fully developed retrocausal account of quantum mechanics yet. So while it may be possible to make quantum mechanics consistent with Humean supervenience, it is not clear whether this will work, and if it will, whether it is worth the cost to our other intuitions.

8.3 The Case Against Individuals

We have seen that holism is a consequence of all three major interpretations of quantum mechanics, but it isn't strictly speaking a consequence of the quantum phenomena themselves, since it might in principle be possible to account for the phenomena while embracing Humean supervenience. But suppose one of the three major interpretations is correct, and Humean supervenience fails. This means that entangled systems have some properties that cannot be reduced to properties of their parts. And entanglement is ubiquitous: The Schrödinger dynamics generally entangles systems that interact, and since everything in the universe has interacted if you go far enough back (to the big bang), it looks like the state of the universe is a giant entangled state.

If holism is taken for granted, what does this tell us about the composition of physical objects? Ordinarily, we tend to think of physical objects as composed of small particles—a statue is "atoms arranged statuewise" (Merricks, 2001, 3). But Schaffer (2010) has suggested that quantum mechanics turns this mereological priority on its head: Rather than small particles being basic and large objects being composed out of them, the *universe* is the only properly basic entity, and smaller objects, be they statues or particles, are what we get when we partition this object in various ways. The argument for this position is quite straightforward: The properties of the parts of an entangled system do not determine the properties of the whole, but the properties of the whole determine the properties of the parts. So for an entangled state, the whole must be metaphysically prior to the parts, since the properties of the parts can be explained in terms of the properties of the whole, but not vice versa. And since everything in the universe is entangled with everything else, the universe is metaphysically prior to all its subsystems.

Schaffer calls this position *priority monism*: At the basic level, there is only one thing. If Schaffer is right, then although there are tables and statues and particles, they are derivative rather than basic. Rather than a fundamental ontology of small, local individuals composing larger objects, there is a fundamental ontology consisting of a single, global individual of which smaller objects are parts. But is he right? Somewhat ironically, the *holism* of quantum mechanics might undermine the argument for priority monism. Schaffer's claim that, for an entangled system, the properties of the parts do not determine the properties of the whole is only correct if one excludes the entangled spin property from the set of properties of the parts. It is true that this property is not a property of particle 1, and neither is it a property of particle 2. But if you have already embraced holism in quantum mechanics, you think that there can be properties of the pair of particles that are not reducible to the properties of the particles individually. Then there is no reason to exclude the entangled spin property as a fundamental property of the pair of particles—although not of either particle individually. In that case, the properties of the parts—individually and collectively—*do* determine the properties of the whole, and there is no reason to give ontological priority to the whole.[8]

Priority monism holds that there are local individuals like quantum particles, but they are not ontologically basic. However, some have concluded on the basis of quantum mechanics that there are no local individuals at all. Consider again the entangled state $|S\rangle$. Is there any property that distinguishes the two particles referred to in this state—any property that particle 1 has but particle 2 lacks? The entangled spin property is no help here, as it is entirely symmetric with respect to the two particles.[9] And one can arrange that the two particles have exactly the same state in terms of position, energy, and so on. For example, the two electrons in a helium atom can have exactly the same energy, a wave function that is spread in exactly the same way around the atomic nucleus, and spins entangled as in state $|S\rangle$ (Ladyman & Ross, 2007, 135). For such a pair of particles, it seems that there is no property at all that distinguishes them.

So why think that there are really two particles here? According to Leibniz's principle of the identity of indiscernibles (PII), distinct individuals must have some property that distinguishes them.[10] If we adopt the PII, then there are not really two distinct particles here; there is just a quantum system with a precise energy, a fuzzy position, and a complex spin property described by $|S\rangle$. On the other hand, we could insist that there are two particles here, even though they are not individuated by any of their observable properties; this requires us to reject the PII.

As French (1989, 445) notes, these two options amount to a kind of metaphysical underdetermination. The phenomena themselves do not tell us whether there are particles and the PII is false, or there are no particles and the PII is true. One might think that saving the PII is reason enough to reject particles. Furthermore, the individuality of quantum particles would have to be entirely beyond empirical reach, and it is not clear that we can make sense of such things, and even if we can make sense of them, it is hard to see what reason there could be to believe in them. That is, the PII reflects a widely held methodological principle against postulating undetectable things (Huggett, 1999, 165), but without them there can be no quantum individuals.

As Ladyman and Ross (2007, 136) point out, none of these purely philosophical considerations is conclusive. The PII is not sacrosanct, and undetectable entities can be defended on nonempirical grounds. In general, if there is no empirical reason to think that there are quantum individuals, then by the same token, there is no empirical reason to think that there are none. But rather than retreating to agnosticism, one kind of principled response to metaphysical underdetermination is that we are demanding too much of our metaphysics. Instead, we should adopt weaker metaphysical commitments according to which the underdetermination doesn't arise.

This is the appeal of ontic structural realism (OSR) as a metaphysical strategy (Ladyman, 1998). The idea is that our ontological commitments should extend to the structure represented by a theory, but not to the nature of the entities instantiating that structure. In the present case, we should be committed to the structure implicit in state $|S\rangle$, and that is as far as our ontological commitments should go. Note that this structure is fundamentally relational: It tells us the correlation *between* the spins, but not the values of the spins themselves. Hence, the basic OSR strategy, when applied to quantum mechanics, is to regard quantum ontology as a structure of relations (French & Ladyman, 2003).

On this view, the basic ontology presented to us by quantum mechanics is one of relations, and the question of the individuality or otherwise of the relata doesn't arise, since the relata are no part of the fundamental ontology. That is, while it might be convenient at times to describe a system in terms of particles, in cases where this is problematic (e.g., the helium atom), one can simply forgo the particle description. If this is coherent, then the question of the truth of the PII is moot and the underdetermination worry is shelved. Interpretations of quantum mechanics that take it to describe relations rather than individuals have been defended by Rovelli

(1996), Mermin (1998), and French and Ladyman (2003). Ladyman and Ross (2007), in particular, present quantum mechanics as the centerpiece of a general argument that there are no *things*—no individuals at the basic physical level.

There are two different kinds of concern one might have regarding this conclusion—its coherence and its motivation. Regarding coherence, you might wonder whether an ontology of relations can possibly be coherent without a commitment to the individuals related by those relations. How could state $|S\rangle$ represent a relation of oppositeness between two spins unless there are two spins between which the relation holds? French and Ladyman (2003) suggest that the way to address this worry is to reverse the usual logical priority we assume between relations and individuals. That is, rather than regarding individuals as fundamental and relations as holding between them, we should regard relations as fundamental and individuals as nodes in a relational structure. This might be conceptually unfamiliar, but it is arguably not incoherent.

Regarding motivation, you might be concerned about the appeal to OSR as a way out of metaphysical underdetermination. OSR is itself a metaphysical doctrine, and itself subject to metaphysical underdetermination; it is obviously not an empirical matter whether or not there are entities beyond empirical reach. So using OSR to overcome metaphysical underdetermination seems to beg the question. Nor can one insist that OSR is mandated by quantum physics, since as we have seen, OSR acts as a premise in overcoming the empirical deadlock between relational quantum mechanics and quantum mechanics with individuals. Perhaps metaphysical underdetermination between quantum mechanics with and without local individuals is inevitable (Frigg & Votsis, 2011, 268). That is, an ontology of relations may be an interesting possibility that is suggested by the structure of quantum mechanics, but it is much harder to show that it is mandated by quantum mechanics.[11]

8.4 Individuals Reconsidered

The previous section treats quantum mechanics at a fairly general level, and at that level the arguments against local individuals are somewhat inconclusive. But we have seen on previous occasions that general arguments about the consequences of quantum mechanics often look very different through the lens of the various available interpretations of the theory. So let us take them in turn.

Hidden variable theories provide the most obvious place to look for individuals in quantum mechanics. In fact, retrocausal hidden variable theories, as we saw earlier, do not require any holism at all, and in principle all quantum phenomena could be explicable in such a theory using only properties of individual particles. There is no motivation to reject individuals here. Bohm's theory does entail holism, but nevertheless the Bohmian particles, like classical particles, can always be individuated in terms of their positions. In state $|S\rangle$, even if the distribution of wave function amplitude is exactly the same for each particle, the two particles have distinct positions within the wave packet. So even though Bohm's theory is holistic, in the sense that entangled systems have *some* properties that are irreducible to the properties of their parts, nevertheless the parts retain some individual properties, and there is no reason to adopt a fundamentally relational ontology.[12]

Individuals are harder to discern in many-worlds and GRW. The representation of the two electrons in a helium atom in these theories is entirely symmetric between the two "particles"—or more precisely, entirely symmetric between the three dimensions associated with particle 1 and the three dimensions associated with particle 2. If the two particles are individuable, it cannot be on the basis of their properties while in this state. Perhaps they could be individuated via their behavior later on; for example, perhaps particle 1 later undergoes a GRW collapse and particle 2 does not, or perhaps particle 1 later ends up in a many-worlds branch in which its position is measured and particle 2 does not. But many (perhaps most) helium atoms will never be subject to this kind of "symmetry-breaking" interaction.

So it looks like there may well fail to be individuable particles in GRW or many-worlds. And this should come as no surprise, since the theories certainly don't appear to describe particles at the micro-level. In each theory, individual objects emerge at the macro-level (relative to a branch in many-worlds). But it doesn't obviously follow that the appropriate fundamental ontology for GRW or many-worlds is a relational one. The most obvious ontology is a field ontology: Perhaps the amplitude and phase of the wave function are defined at each point in configuration space, or perhaps the value of a mass density or some other kind of field is defined at each point of ordinary three-dimensional space (see Chapter 7). Whatever the nature of the field, the lack of individuality of quantum particles is manifest in the fact that some wave functions, like the helium atom wave function, are entirely symmetric in the coordinates of two or more of the "particles" involved.

However the quantum field is envisioned, though, it reintroduces an ontology of individuals—of individual points rather than individual particles. The individuation of points of space—configuration space or ordinary three-dimensional space—itself raises some tricky philosophical issues, but this debate would take us too far into the physics of space-time for the current volume. At least as far as quantum mechanics itself goes, there is no reason to reject an ontology of individuals. Similar comments apply to the "flashy" version of GRW (Chapter 5). Recall that according to this version, the fundamental constituents of the world are point-like events corresponding to the center-points of the GRW collapses. Such events are clearly individuals, individuated by their spatial and temporal properties. Again, such individuation may be problematic on other grounds, but there is nothing in quantum mechanics itself that threatens an ontology of individuals.

8.5 Conclusion

As is our usual refrain, quantum mechanics does not conclusively settle the issues surrounding wholes and parts. But it provides some much-needed clarity: Holism and emergence can be made precise using examples from entangled quantum systems. And even if quantum mechanics doesn't definitively answer all our ontological questions, it at least strongly suggests some things. First, unless a retrocausal hidden variable theory can be made to work, it certainly looks like some form of holism is inevitable; entangled systems have properties that cannot be reduced to properties of their parts, and hence Humean supervenience fails. Second, although this holism looks at first like it might undermine the existence of local individuals in quantum mechanics, in fact the existence of local individuals is secure under each of the major interpretations.

9| Six Quantum Worlds

IN 1986, JOHN BELL GAVE a talk at a Nobel Symposium in Sweden entitled "Six Possible Worlds of Quantum Mechanics" (Bell 2004, 181). In it he summarizes the ways the world could be such that quantum mechanics is true of it—possible ontologies of the physical world at the quantum level of description. This seems like exactly the right way to finish this book. As we have seen over and over, the foundations of quantum mechanics are so contested that it yields few definitive answers to metaphysical questions. So the best we can do is to lay out the various possibilities.

Strictly speaking, some of the entries on Bell's list aren't really possible worlds at all. The first item on his list—the pragmatic approach—eschews all attempts to describe the world and concentrates on how best to apply quantum mechanics in the practice of physics. There is no description of a possible *world* here. And two other items on his list—Bohr's "complementarity" approach and Wigner's dualistic approach—are attempts to describe the world, but (for different reasons) fall short of being *possible*. Complementarity, as Bell himself notes, seems to apply contradictory descriptions to a physical system without explaining clearly how the contradiction is to be resolved (2004, 190). And Wigner's view, according to which wave function collapse is triggered by the intervention of human consciousness, requires a deeply problematic interactionist dualism.[1]

The remaining three items on Bell's list are familiar: spontaneous collapse theories, hidden variable theories, and many-worlds theories. The possible ontologies he sketches here are the ones we have been exploring in detail in the foregoing chapters. But there are not exactly three possibilities here: As we have seen, these general interpretive strategies can be developed in various ways, and how many distinct ontologies there are exactly will depend on how you choose to individuate them. So the following is a list of six quantum worlds, but the number is really indeterminate.

Let's start with Bohm's theory. Bohm's theory presents us with a clear, dual ontology: There are particles, and there is a wave-like field that pushes them around. The particles follow continuous trajectories, and the dynamical laws that govern both the wave and the particles are entirely deterministic. The particles themselves have very few intrinsic properties—perhaps just positions—and other properties are better conceived of as properties of the wave. For properties other than position, the value of a particular determinable property can be indeterminate; there might be no fact of the matter about the spin of a particle, for example. Furthermore, some properties of a system of several particles can only be ascribed to the system as a whole and cannot be reduced to the properties of the parts. Finally, the causal law governing the motion of the particles is nonlocal, in that the motion of a particle can depend on what is going on right now in distant parts of the system. It is this last feature that might be the downfall of Bohm's theory, as it puts it in conflict with special relativity.

Second, there are variant hidden variable theories. The most interesting of these from an ontological point of view are "particle only" hidden variable theories, which do without the wave as part of the ontology. These are somewhat tentative; there is no fully detailed theory here. One way to do without the wave is to try to interpret the wave function as describing a law rather than an entity. Another way is to allow causal influences in both temporal directions—from later events to earlier as well as vice versa. The latter strategy is particularly interesting. Again, the particle trajectories are continuous, but now it looks like the laws that govern them can be fully local as well. The particles themselves carry the properties of the system—particles are not merely bare positions—and there seems to be no reason why these properties should not always be fully determinate. Furthermore, it looks like all the properties of compound systems ought in principle to be reducible to the properties of the particles; there is no need for holism. One might also wonder whether retrocausal theories are likely to be deterministic, but here it is hard to say, not only because we don't have the causal laws in front of us, but also because it is hard to define determinism in a retrocausal context.[2] The main challenge for this approach is the formulation of a precise and coherent theory.

Third comes the GRW theory. The ontology here is wave-like, and the dynamical law that governs the wave is indeterministic, since it includes collapse events at random times centered on random locations. The properties of things can sometimes be indeterminate—even their positions, and even the positions of observable things like photons. But in general the GRW collapse process ensures that the positions of macroscopic objects are determinate, and collapses in our brain mean that we aren't aware of the indeterminate positions of photons. Furthermore, the GRW theory exhibits holism: The properties of a compound system are sometimes irreducible to the properties of its parts. Finally, the GRW collapse process is causally nonlocal: a collapse in one location has instantaneous physical effects in other locations. As in Bohm's theory, it is the nonlocality of the GRW theory that is most problematic.

Fourth, there are variants of the GRW theory that add additional primitive ontology, most notably a mass density distribution over space. This doesn't change any of the features just noted: The theory is still indeterministic, holistic, and nonlocal. What it achieves is an ontology that undeniably lives in three-dimensional space, unlike the wave function, which requires a much higher dimensional space. But since the wave function (I have argued) can itself be interpreted as describing a three-dimensional world, it is not clear that this extra layer of ontology buys us anything.

Fifth is the "flashy" spontaneous collapse theory. The ontology here is of discrete, point-like events in space-time. So the ontology is intermittent: A macroscopic object is a swarm of points spread over a continuous space-time trajectory, but a microscopic "object" may not correspond to any ontology at all. Instead, the microscopic system has to be reconceived in terms of correlations between the behavior of the macroscopic objects involved in preparing it and the behavior of macroscopic objects involved in detecting it. The flashy ontology does nothing to address indeterminacy, holism or indeterminism, but arguably allows the spontaneous collapse approach to be reconciled with special relativity. The cost of doing this is that there are cases in which the direction of causation between two events is indeterminate: Which is the cause and which is the effect depends on the choice of coordinates.

The many-worlds theory completes our list. The ontology here is, again, wave-like, but in this case causal explanations also appeal to the self-location of observers within the branching structure of the waves. The causation involved here is arguably entirely local. The properties of things are generally only determinate when considered from the perspective of a particular branch, and even then the properties of microscopic systems are often indeterminate. Furthermore, the properties exhibit holism: Properties of compound systems can fail to be reducible to the properties of their parts.

The wave evolution in the many-worlds theory is deterministic at the global level, but it offers at least the appearance of indeterminism at the level of the individual branch. The main problem facing the many-worlds theory concerns whether a theory in which every possible outcome of an experiment actually occurs (in some branch) can be made to yield the familiar probabilities of quantum mechanics.

There are other ontological positions that have been considered over the course of the previous chapters, but which are not obviously a consequence of any of the above theories just mentioned. These include the position that the world has a very high spatial dimensionality (i.e., it is not really three-dimensional), the position that the whole universe is metaphysically prior to its parts (priority monism), and the position that there are no local individuals (i.e., that the fundamental ontology is one of relations). This is not to say that these positions have been ruled out, just that quantum mechanics gives us no special reason to endorse them.

So very little can be concluded unconditionally on the basis of quantum mechanics: Metaphysical claims of the form "Quantum mechanics shows that ... " need to be treated very carefully, and in their full generality are likely to be false. However, this doesn't mean that thinking about quantum ontology is a useless exercise. The empirically informed debate over ontological issues generated by quantum mechanics is often quite unlike the standard debates over these issues, and the range of possibilities entertained is often different, too. Even if quantum mechanics doesn't settle many ontological questions, it shifts the debate in interesting and fruitful ways. Furthermore, the metaphysical consequences of the various interpretations of quantum mechanics may, in some cases, reflect back on the tenability of those interpretations. If an interpretation cannot yield a coherent metaphysical picture of the world, then it cannot be regarded as an adequate descriptive theory.

Quantum mechanics is fascinating and frustrating. Its phenomena are astonishingly difficult to fit into any coherent ontological framework. The frameworks we end up with are fascinatingly revisionary but also frustratingly problematic. The best we can say is that not *everything* in our received classical worldview can be right.

NOTES

Introduction

1. The *way* in which classical mechanics is a limiting case of quantum mechanics is problematic, and a symptom of the more general problem of interpreting quantum mechanics to be explored in this book. Nevertheless, most would agree that classical mechanics must be a limiting case of quantum mechanics in *some* sense.

2. See Barrett (2008) for one way this might go.

Chapter 1

1. Of course, this initial impression might be wrong: It might be possible to retain our classical metaphysics with a little creativity. The goal of this book is precisely to assess our options. It will turn out that there are various ways of recovering part of our classical metaphysical picture, but always at the expense of some other part.

2. In fact, things are a little more complicated, since the charge of the electron also interacts directly with the external magnetic field, but the idealization here will do no harm.

3. The name "spin-1/2" comes from the fact that the angular momentum associated with the spin of these particles is $\hbar/2$, where $\hbar$ is Planck's constant divided by 2π.

4. In general, these numbers are complex, but the examples in this section have been chosen to keep all the numbers real. Often vectors are written as vertical lists rather than horizontal lists, but I will write them horizontally for ease of display.

5. The notation $|a|^2$ means a multiplied by its complex conjugate, since in general a and b can be complex. In the current context where all the numbers are real, $|a|^2$ is the same as a^2.

6. Precisely, $(1,0)$ can be represented as the weighted sum $a(a,b) - b(-b,a)$, since $a(a,b) - b(-b,a) = (a^2, ab) - (-b^2, ab) = (a^2+b^2, ab-ab) = (1,0)$, using $a^2 + b^2 = 1$ for the last step.

7. In cases where the vector elements are complex, the projection of (a_1, b_1) onto (a_2, b_2) is $a_1 a_2^* + b_1 b_2^*$, where a^* is the complex conjugate of a.

8. At least, position is treated as continuous in standard quantum mechanics, even if it may ultimately turn out that spatial position, like spin, is quantized.

9. The sense in which the continuous case is a *limit* needs to be treated with some care, though. Matrix mechanics is defined on a separable vector space—one in which the number of basic vectors is countable. The continuous case doesn't correspond to a separable vector space.

10. As noted earlier, matrix machanics is defined on a separable Hilbert space, so the number of basic states must be countable.

11. Physicists frequently refer to mathematical entities called *delta functions*, written $\delta(x)$, that are apparently ascribed these impossible properties. However, this is just a form of innocuous mathematical shorthand, useful in simplifying certain integral expressions.

12. Of course, one might deny that electrons have precise positions. The possibility of this kind of indeterminacy in the world is explored in Chapter 4.

13. If you think that a measurement rule is *not* part of standard (classical) physical explanations, you have a point. This point is explored in Chapter 3.

14. Note that I am using "interpreting" in a broad sense, to include supplementing or even replacing quantum mechanics as a physical theory. Some may wish to divide the "genuine interpretations"—those that work with quantum mechanics as it stands—from the more revisionary projects. But I will stick with the broad usage, which has become somewhat canonical in the literature.

Chapter 2

1. The term "incomplete" needs to be understood in a particular way. The claim is not that quantum mechanics fails to provide an "ultimate" description of the most fundamental constituents of matter; that is certainly true, as discussed in the Introduction. Rather, the claim is that quantum mechanics fails to adequately describe the world at the level of description appropriate for explaining canonical quantum phenomena such as interference and entanglement.

2. Consider the general basis states $|\uparrow\rangle = (a,b)$ and $|\downarrow\rangle = (-b,a)$, where $a^2 + b^2 = 1$. Then we can write $|\uparrow_z\rangle = a|\uparrow\rangle - b|\downarrow\rangle$ and $|\downarrow_z\rangle = b|\uparrow\rangle + a|\downarrow\rangle$, as shown in Chapter 1. Substituting these expressions for $|\uparrow_z\rangle$ and $|\downarrow_z\rangle$ in state $|S\rangle$ gives $|S\rangle = \frac{1}{\sqrt{2}}[(a|\uparrow\rangle - b|\downarrow\rangle)(b|\uparrow\rangle + a|\downarrow\rangle) - (b|\uparrow\rangle + a|\downarrow\rangle)(a|\uparrow\rangle - b|\downarrow\rangle)] = \frac{1}{\sqrt{2}}(|\uparrow\rangle|\downarrow\rangle - |\downarrow\rangle|\uparrow\rangle)$; that is, it has the same form as when written in the z-spin basis.

3. In fact, as we shall see in Chapter 3, the collapse postulate can't be right even for localized systems.

4. And, of course, the same goes for the first electron, since we could measure the spin of the second electron first. And the same goes for any other direction we might choose.

5. My interpretation of Bohr here follows Howard (1994) and Halvorson and Clifton (2002). As Howard (1994, 204) notes, it is not easy to discern Bohr's

intentions from his words, and to this extent the interpretation is something of a reconstruction. But nevertheless, these authors make a good case that the reconstruction is a charitable one.

6. Bohr's own exposition, like that of the EPR article, concerns the position and momentum of a particle. I have translated the discussion to the simpler context of spin properties.

7. Furthermore, complementarity has a straightforward mathematical analog in quantum theory itself. Every physical property of a system corresponds to a mathematical operator on the vector space (or the space of wavefunctions) used to represent the system. Sometimes the operators corresponding to two different properties *commute*, meaning that the order in which the operators are applied makes no difference. That is, operators $\widehat{A}$ and $\widehat{B}$ commute iff for any state $|\psi\rangle$, $\widehat{AB}|\psi\rangle = \widehat{BA}|\psi\rangle$. Bohr's complementary properties are properties for which the corresponding operators fail to commute.

8. One such scheme is to assign the properties in one of the rows 1–8, picked at random with equal probabilities. If a row from 2 through 7 is chosen, then stipulate that the measurement directions are such that the results always agree. For example, for row 2 stipulate that the measurement directions are vz, wz, zv, or zw, again picked at random with equal probabilities. Hence, for rows 2–7 (75% of the time) the results agree, and for rows 1 and 8 (25% of the time) the results disagree.

9. For example, if the first electron will be measured along v and the second along w, then assign the properties in row 1, 2, 7, or 8 (at random with equal probabilities) 25% of the time, since the v-spin of electron 1 and the w-spin of electron 2 differ in these rows. The remaining 75% of the time, assign the properties in row 3, 4, 5, or 6 (at random with equal probabilities), since the v-spin of electron 1 and the w-spin of electron 2 agree in these rows. Similar, recipes apply to the other possible measurement directions.

10. If the first electron is measured to be spin-up along v, then assign the properties in rows 1–4 (at random) to the electrons with probabilities 1/16, 3/16, 3/16, and 9/16, respectively, so that whether the second electron is measured in the w-direction or the z-direction, the result is spin-up 3/4 of the time. Similar, recipes apply to the other possible outcomes for the first electron.

11. My presentation of Bell's theorem follows Mermin (1981).

12. As Mermin (1993) points out, a version of this theorem was proved by Bell (1966), but it is nevertheless usually credited to Kochen and Specker (1967).

13. A simpler proof can be given in a four-dimensional vector space; see, for example, Mermin (1993).

14. In fact, as Bell (1966) notes, many physicists were convinced that von Neumann (1932) had already shown the impossibility of hidden variable theories. Bell goes on to show the inadequacy of this proof.

15. For a balanced discussion of this form of argument, see Dickson (1995).

16. The measurement postulate cannot describe this process because such a process would involve instantaneous action at a distance, as noted earlier. The various failings of the measurement postulate as a piece of physics are explored further in Chapter 3.

17. I use position here just as an example, but it is an important example, as position is singled out as a special property by several realist interpretations of quantum mechanics, in particular Bohm's theory and the GRW theory. These will be properly introduced in the following chapter.

18. The same goes for any other measurable quantities apart from position. One might think that this makes properties like spin contextual, and hence means that Bohmian approaches violate independence. However, since there are no spin properties according to this approach, there is no real violation of independence here: There are only position properties, and these are not contextual.

19. The many-worlds approach differs from the spontaneous collapse approach and the Bohmian approach in not singling out position (or any other property) as special. A system whose state is not a position eigenstate has no preexisting position properties either.

20. This, of course, is prima facie in conflict with special relativity. This problem is investigated in Chapter 5. It is interesting to note that the standard theory of quantum mechanics introduced in Chapter 1 also involves this kind of action at a distance, insofar as the measurement postulate is understood as a physical collapse process.

21. Of course, you may be convinced by the *global* arguments against scientific realism mentioned at the beginning of this chapter. But even so, unless you are a verificationist about meaning, you will still want to interpret scientific theories as literally descriptive of the world, even if you don't believe that description (van Fraassen, 1980, 11). In that case, even a scientific antirealist might regard quantum mechanics as an inadequate theory that is in need of "completion" in one of the ways described in the next chapter (van Fraassen, 1991, 273).

Chapter 3

1. There are empirical differences between some of the versions of quantum mechanics described later, but they are subtle and hard to investigate; there is no reason to think they will be resolved soon.

2. Since the GRW collapse occurs to a narrow Gaussian rather than to a precise point, the recipe for constructing the probability distribution is actually a little more complicated: It is the squared precollapse wavefunction weighted by the Gaussian hit function. That is, the probability distribution over the possible hit-points x_i is $\int |\psi(x) G(x, x_i)|^2 \, dx$, where $\psi(x)$ is the prehit wavefunction, $G(x, x_i)$ is the three-dimensional Gaussian hit function centered on point x_i, and the amplitude of G is chosen such that the integral of the probability distribution over x_i is 1.

3. That is, the second wavefunction peak—the one centered on B for each particle—doesn't disappear, it just becomes very small. It doesn't show up on the right-hand side of Figure 3.3 because only regions of large wave amplitude are shaded.

4. But note that in Chapter 4 I explore the possibility that we cannot tell whether objects have well-defined locations or not.

5. More carefully: the interaction between the properties in different sets is negligible unless the two terms interfere, and interference is effectively impossible to arrange for macroscopic systems. This is discussed in more detail in Chapter 4.

6. Everett developed his interpretation of quantum mechanics as his Ph.D. dissertation, but then left academia, leaving many unanswered questions about how he understood his own theory. Recently discovered documents shed more light on Everett's own views; see Everett, Barrett, and Byrne (2012).

7. Talk of *worlds* in this context probably originates with DeWitt (1970).

8. For an exposition of the sense in which Everett himself understood his theory to be empirically adequate, see Barrett (2015).

9. See Barrett (2011). Barrett suggests that Everett didn't care much about the choice of language.

10. Some Bohmians argue that the wave function can be interpreted as a *law*; this possibility is discussed in Chapter 7.

11. It is not as clear for the GRW theory as for Bohm's theory that both patterns remain: It follows straightforwardly if the dynamical laws are always linear, but the GRW collapse process is a nonlinear addition to the linear Schrödinger dynamics. See Wallace (2014) for a discussion.

12. Although if the wave function is a field, then as noted earlier, it is a field defined on a high-dimensional space.

13. See also Ney (2013 177). Ney argues against Dennett's criterion on the grounds that typical reductionist strategies rely on a specific conception of the base ontology. That is, while a particular pattern *in the spatial arrangement of small parts* might be sufficient for the existence of a cat, it doesn't follow that the pattern itself is sufficient. In that case, it is possible that the reduction of cats to patterns in the Bohmian particles succeeds while their reduction to patterns in the wave function fails.

14. D. Lewis (1976) devises this account to deal with personal identity in cases of Parfittian fission. Saunders and Wallace (2008) adapt it to the description of objects in Everettian quantum mechanics.

15. The reason such experiments are so hard is that any unanticipated interaction between the object in question and its environment makes it impossible to determine whether the object is in a superposition state or not, and it is extremely difficult to control the interactions between a macroscopic object and its surroundings. See Albert (1992 p .88).

Chapter 4

1. The cloud-like electron can be analyzed in a similar way. There are discrete location eigenstates in which precisely all the wavefunction amplitude is contained in a particular small region, and in these states the electron has a determinate location. But for all other states, including the cloud-like one pictured, the electron simply lacks a determinate location. This analysis makes no mention of a continuum of properties. But it may also be the case that the analysis in terms of familiar vagueness and a continuum of properties is appropriate here; I return to this example in the conclusion of the chapter.

2. Later we will introduce alternatives to the strict link, the fuzzy link and the vague link, according to which there *is* vagueness in the quantum world, although not of the familiar kind.

3. The projection is given by the coefficient on the relevant eigenstate when the state is written as a superposition of the eigenstates of a particular observable. See Chapter 1 for the details.

4. A many worlder might argue that this latter interpretation is unavaliable, as the many-worlds interpretation of the state is forced on us by the patterns each term contains. This form of argument was considered in Chapter 3 and found to depend on somewhat controversial claims about the existence of probabilities in Everettian quantum mechanics.

5. Barrett brings up the example of the apparent bending of a spoon when its handle protrudes through the surface of a glass of water as a case in which insisting on the way things seem would be a bad strategy (1999, 112). A better example for present purposes would be a case where we are even wrong about our own experience. For example, the physiology of perception entails that our visual field outside a narrow "foveal region" is much less distinct than we take it to be; a case can be made that we should regard this theory as undermining our apparent introspective evidence concerning the distinctness of our peripheral vision (Schwitzgebel, 2008).

6. What's more, the proponent of the bare theory can deny that any such objection can be made, on physical grounds. Suppose the physical state of the world is described by a complicated superposition of terms, some of which describe a coffee mug on my desk and some of which do not. Then my mental state is described by a superposition of terms, too, some of which describe me as seeing the mug and some of which do not. So when I try to utter the objection, the physical state of my vocal apparatus will be described by a superposition of terms, some of which describe me as asserting that I see the mug and some of which do not. So there is no determinate utterance that I made, although it seems to me that I made a determinate utterance. Somewhat exasperatingly, the bare theorist can insist that I did not utter a determinate objection to the bare theory, even though it seems to both of us that I did!

7. Barrett (2015) investigates the extent to which other interpretations of Everett might be empirically incoherent.

8. This assumes that a theory of meaning must appeal to features of the world. I don't have an argument that this must be the case, except by induction on the theories just enumerated.

9. Eric Schwitzgebel suggested to me that perhaps under the bare theory the meaning of each statement is indeterminate, although it seems to any linguistic beings that the statement has a determinate meaning. Such a theory might even be true in some vacuous sense, since it says nothing. But then the bare theory still lacks all content, and a contentless theory really isn't a theory at all.

10. That is, unless one wants to follow Williamson (1994) in holding that there is a precise boundary, but that it is unknowable.

11. Note that the proposal embodied in the vague link is not the same as higher order vagueness: It is not that determinacy is an all-or-nothing thing, but that it can be indeterminate whether a property is determinate or not (e.g., Hyde, 1994). The reason is that higher order vagueness doesn't solve the problem we are considering here. One could try saying that a system only determinately possesses a determinate property when its state is the eigenstate. But this would entail that the systems we

interact with do not determinately possess the determinate properties we ascribe to them. This doesn't look like progress.

12. Of course, one might conceive of a kind of continuum in the classical case, too, by considering moving one particle at a time from the desk to the drawer, but the situation here is quite different: The continuum doesn't involve decomposing the mug into parts at all.

13. One can imagine creatures who have different brain structures from ours and store certain simple perceptual states in the positions of just a few particles (Albert 1992, 108). For such creatures, not all measurements would be ascribed determinate outcomes by the GRW collapse process, even taking the brain of the observer into account (although the creatures wouldn't notice this indeterminacy). But it is not clear to what extent this is an objection to the GRW theory.

14. Roman Frigg has suggested to me that we should leave ordinary language behind: As a revisionary proposal, counting *should be* about individuals rather than ensembles. This works but seems to me to take the counting anomaly too seriously (as discussed later).

15. For an accessible introduction to decoherence, see Zurek (1991).

16. For the reasons discussed in Chapter 1, these eigenstates will not be precise position eigenstates for the object, but rather eigenstates for a more coarse-grained variable such as being located in a particular region of space.

17. And even the determinate location of the object used to define the branches is accidental, since we could have chosen some other object (or some other property) as the basis for the branching.

18. It is possible to imagine creatures who encode simple perceptual states in properties other than position, and for such creatures it is conceivable that some measurements might lack determinate outcomes under Bohm's theory (Albert, 1992, 173). But as for the GRW theory, it is not clear whether this is an objection to Bohm's theory.

19. This may be very hard to achieve in practice. But note that a collection of marbles large enough to manifest the counting anomaly is impossible to achieve in practice!

20. But perhaps not impossible. Some Bohmians argue that the wave function can be interpreted as a *law*; this possibility is discussed in Chapter 7.

Chapter 5

1. That is, even in purely classical cases, a measuring instrument might affect the measured system in uncontrollable ways. In that case, we might not be able to describe the causal process giving rise to the measurement result, but it doesn't tell us that we need to give up on the idea of causal processes altogether.

2. It is worth noting, though, that wave function is represented in a multidimensional space, not in ordinary three-dimensional space. Whether this is a problem is taken up in Chapter 7.

3. The following example is from Albert (1992, 156). I learned the kind of diagram used here from Barrett (e.g., 2000).

4. A further criticism sometimes made of Bohm's theory is that the trajectories described by the theory are unrealistic, in the sense that they differ from the actual

trajectories of the particles involved (Englert, Scully, Süssmann, & Walther 1992). But as Barrett (2000) has argued, such arguments do not really show that Bohmian trajectories are unrealistic, but just serve as a further illustration of the causal nonlocality of Bohm's theory.

5. The connection between violations of independence and free will is explored in Chapter 6.

6. As we shall see in Chapter 7, it may be possible to regard the Bohmian wave function as a law rather than as an entity. But in any case, the traditional view for Bohm's theory is that the wave function is a dynamical entity, and certainly it is a dynamical entity in GRW and Everett.

7. However, we do have reason to think that Everett's theory exhibits a kind of *holism* that won't go away. That is, we have reason to think that the properties of spatially disjoint systems aren't reducible to the properties of their spatially localized parts. But holism doesn't entail causal nonlocality; see Wallace and Timpson (2010) for a discussion. Holism is taken up in Chapter 8.

8. A single particle can decay into two particles, of course, but that is not what is going on here. The initial wave packet represents a particle with a certain mass and charge—those of an electron, say—and each of the wave packets after the split represents a particle with exactly the same mass and charge.

9. See the discussion in Chapter 4 about when exactly this collapse process occurs. Collision with a fluorescent screen is not enough to trigger a collapse, so the many nascent branches will persist until a sufficient number of particles become involved in the process.

10. In fact, since the wave packet has small "tails" spreading to infinity that are affected by the collapse event, the measurement of particle 2 has instantaneous effects everywhere in space.

11. Maudlin (1994, 198) also argues that the causal stories told by the transactional theory produce inconsistent causal loops. Strategies for trying to solve this problem can be found in Lewis (2013b) and Kastner (2013).

12. This surface is a surface of fixed space-time interval from the previous flash—that is, $x^2 + y^2 + z^2 - c^2 t^2 = S^2$, where c is the speed of light and S is a constant. See Tumulka (2006b, 347) for a diagram.

13. Presumably for this choice of coordinates there is no fact of the matter about which is the cause and which is the effect.

Chapter 6

1. The outcome also depends on the orientation of the measuring devices—but this is also determined by the positions of all the Bohmian particles in the vicinity.

2. Even if your eyes are open, there will be a brief period of time after the spin-up and spin-down branches have separated but before your visual system has had a chance to register the outcome. However, I doubt that it is appropriate to attribute epistemic states like *uncertainty* over such a brief period of time.

3. This is an example of an essential indexical in Perry's (1979) sense; the "I" in "I wonder what I will see when I open my eyes" cannot be replaced with any nonindexical expression without destroying the ground of the uncertainty. There is

a good deal of literature on the connection between Everettian uncertainty and other kinds of self-location uncertainty, especially in Sleeping Beauty cases, but exploring it would take us too far afield. See Lewis (2007a, 2009); Bradley (2011); Groisman, Hallakoun, and Vaidman (2014).

4. If I am a passenger in a car on the road, I can be uncertain about whether *I* am going to Sebring. But the analogy is between the *road* and the Everettian person.

5. D. Lewis (2004, 15) suggests that we adopt as a physical postulate that people are distributed over branches according to the Born rule. He realizes (more than most!) that such people cannot be individuated by future branching, so he simply stipulates that branches contain mutiple copies of a person. As he recognizes, this is essentially a form of Albert and Loewer's (1988) many-minds theory and suffers from its drawbacks.

6. The original version of this proof can be found in Everett (1957). See Wallace (2012, 125) for a discussion of other versions. As Wallace notes, Everett's proof is just the proof of the law of large numbers in probability theory translated into the language of quantum mechanics. (One version of) the law of large numbers says that for a repeated chance event, the probability of getting relative frequencies of outcomes that are close to their probabilities tends to 1 as the number of repetitions tends to infinity.

7. Note that the *number* of branches exhibiting a particular relative frequency is independent of the amplitudes a and b. For example, after two repetitions, there are two branches containing 50% spin-up results, one branch containing 100%, and one branch containing 0%, regardless of the values of a and b. As the number of repetitions tends to infinity, the proportion of branches containing a relative frequency of spin-up significantly different from 50% tends to zero (by the law of large numbers). So unless $|a|^2 = |b|^2 = 1/2$, the proportion of branches in which the frequency of spin-up results is close to $|a|^2$ tends to zero.

8. One might worry here about the fact that even in the case of more-or-less instantaneous death mechanisms, there will still be a fraction of a second after the branching event and before the mechanism kills you. But it is plausible, at least, that personhood is a temporally coarse-grained notion, so that it makes no sense to talk of a postbranching person lasting a fraction of a second.

9. Wallace does endorse the existence of uncertainty in Everettian quantum mechanics, as noted earlier, and he does analyze uncertainty in terms of Lewisian temporally extended persons. But for Wallace, the explanatory order precludes drawing the immortality conclusion: The decision-theoretic argument for the Born rule is prior to any particular analysis of uncertainty.

10. Fifty-nine percent of philosophers responding to the 2009 PhilPapers survey endorsed compatilbilism, as against fourteen percent for libertarianism. See Bourget and Chalmers (2014).

11. See also Bell's comments on free will in Davies and Brown (1986, 47). However, it is not clear to what extent Bell was worrying about free will per se rather than about the requirement of a vast hidden causal order discussed later. For elaboration, see Price (1996, 237) and P. Lewis (2006a).

12. Bell said of the retrocausal approach that "when I try to think of it I lapse quickly into fatalism" (Price, 1996, 241). But this looks like a mistake: A slight restriction on compatibilist free will is far from fatalism.

Chapter 7

1. Strictly, since the wave function is a *function*, it is mathematical rather than physical. That is, strictly speaking the wave function itself is a mathematical representation of an (unnamed) physical state of affairs. Maudlin (2013, 129) suggests the name *quantum state* for the physical state of affairs, but in Chapter 1 I chose to use this term as a name for the mathematical representation, reserving *physical state* for the state of affairs represented. However, I am less concerned than Maudlin about the possibility of ambiguity, and it is traditional to use the term "wave function" for both the mathematical representation and the physical object represented. Where it might be important, I specify which I mean.

2. At least, at the relevant level of description. As noted earlier, string theories postulate nine or ten spatial dimensions, but only three are relevant to the observable world. This possibility is untouched by the arguments here.

3. The term "primitive ontology" is Allori's; Maudlin prefers "primary ontology."

4. Precisely, the mass density is described by a function $\mathcal{M}(\mathbf{r})$ over three-dimensional space $\mathbf{r}$ defined by $\mathcal{M}(\mathbf{r}) = \langle \psi | M(\mathbf{r}) | \psi \rangle$, where $|\psi\rangle$ is the quantum state and M is the mass density operator for region $\mathbf{r}$ (Ghirardi, Grassi, & Benatti, 1995, 16). Standard quantum mechanics interprets this quantity as the expectation value for a mass density measurement in this region; Ghirardi, Grassi, and Benatti interpret it as the *actual* mass density in the region.

5. This is the so-called *problem of time* in quantum cosmology: It is regarded as a problem because it seems to imply that nothing changes. However, in a Bohmian theory, change is not ruled out even with a static wave function because the *particles* can move over time.

6. Although as Callender (2015) notes, time-evolving laws should perhaps not be ruled out a priori; perhaps quantum mechanics is telling us something surprising about the nature of laws.

7. North (2013) argues that while the world may be three-dimensional under such interpretations, it is not *fundamentally* three-dimensional. If we disagree, I suspect it is over whether spatial three-dimensionality resides at the fundamental physical level, or whether, like solidity, it is closer to the phenomena.

Chapter 8

1. The measurement event is not strictly a *common* cause in this case, as it is intermediate in the causal chain between the choice of measurement type made by the experimenter and the earlier properties of the particle measured. But in any case, Hawthorne and Silberstein's assumption fails.

2. Silberstein (2015) recognizes that retrocausal accounts need not violate Humean supervenience.

3. In fact, it is much worse than that, because entanglement can hold between large numbers of particles and to varying degrees.

4. You could of course reduce the relation "taller than" to the intrinsic properties "5 feet 8" and "5 feet 6," but that won't work in the quantum case.

5. Wallace and Timpson (2010, 712) also conclude that Everett's theory violates Humean supervenience.

6. I assume a retrocausal theory with a particle ontology here (e.g., Price, 2012). But a retrocausal theory with a wave ontology (e.g., Wharton, 2010) would do, too; in this case localized wave packets would be the bearers of the properties.

7. There is a difference between the retrocausal case and the classical case. In the retrocausal case, given the presence of both forward-in-time and backward-in-time causal laws, property ascriptions *at a single time* can violate the causal laws. But this difference doesn't appear to be significant.

8. Similar arguments can be found in Bohn (2012) and Calosi (2014). But as Schaffer quite appropriately pointed out to me, even if priority monism is not mandated by quantum mechanics, it is significant that it is even among the tied-for-best interpretations, alongside some form of collective property possession by fundamental individuals.

9. Strictly, state $|S\rangle$ is antisymmetric, as exchanging the two particle indices introduces a minus sign. But in quantum mechanics the negative of a state is taken to represent exactly the same physical situation as the original state, so the antisymmetry cannot be used to distinguish the two particles.

10. The PII admits of various interpretations. If one excludes spatial properties, then classical particles, too, can have all their properties in common. However, if spatial properties are included, then at least prima facie, classical particles are always discernible but quantum particles are not. It is this latter interpretation I assume here. See French and Krause (2006, 152).

11. But as noted earlier regarding priority monism, it is significant that an ontology of relations is even among the main interpretive options.

12. French and Krause (2006, 174) note that the attribution of properties to particles (rather than to the wave function) in Bohm's theory has been challenged, for example, by Brown, Dewdney, and Horton (1995) and by Bedard (1999). But these authors do not challenge the attribution of *positions* to individual Bohmian particles, so their individuation remains unproblematic.

Chapter 9

1. One problem is that it seems utterly mysterious on this view how the ability to collapse wavefunctions could have evolved with the evolution of conscious beings, since evolution is a purely physical process. For further discussion, see Chalmers (1996, 156 and 356).

2. You might want to call retrocausal theories indeterministic because the past of a particle doesn't determine its future. But this criterion seems inappropriate, because the causal influences on a particle from its past don't exhaust the causal influences on it; there are also causal influence from its future. But if you include the future, you risk making determinism trivial: Of course the future of a system determines its future! To make any progress here, we will have to wait to see what the dynamical laws of any worked-out retrocausal theory look like.

REFERENCES

Aharonov, Y., & Vaidman, L. (1990). Properties of a quantum system during the time interval between two measurements. *Physical Review A, 41*, 11–20.

Aicardi, F., Borsellino, A., Ghirardi, G. C., and Grassi, R. (1991). Dynamical models for state-vector reduction: Do they ensure that measurements have outcomes? *Foundations of Physics Letters, 4*, 109–128.

Albert, D. Z. (1992). *Quantum mechanics and experience*. Cambridge, MA: Harvard University Press.

Albert, D. Z. (1996). Elementary quantum metaphysics. In J. Cushing, A. Fine, and S. Goldstein (Eds.), *Bohmian mechanics and quantum theory: An appraisal* (pp. 277–284). Dordrecht, The Netherlands: Kluwer.

Albert, D. Z. (2010). Probability in the Everett picture. In S. Saunders, J. Barrett, A. Kent, & D. Wallace (Eds.), *Many worlds: Everett, quantum theory, and reality* (pp. 355–368). Oxford, UK: Oxford University Press.

Albert, D. Z. (2013). Wave function realism. In A. Ney & D. Z. Albert (Eds.), *The wave function* (pp. 52–57). Oxford, UK: Oxford University Press.

Albert, D. Z., & Barry Loewer, B. (1988). Interpreting the many worlds interpretation. *Synthese, 77*, 195–213.

Albert, D. Z., & Loewer, B. (1996). Tails of Schrödinger's cat. In R. Clifton (Ed.), Perspectives on quantum reality (pp. 81–91). Dordrecht, The Netherlands: Kluwer.

Albert, D. Z., & Vaidman, L. (1989). On a proposed postulate of state-reduction. *Physics Letters A, 139*, 1–4.

Allori, V. (2013). Primitive ontology and the structure of fundamental physical theories. In A. Ney & D. Z. Albert (Eds.), *The wave function* (pp. 58–75). Oxford, UK: Oxford University Press.

Allori, V., Goldstein, S., Tumulka, R., & Zanghì, N. (2011). Many worlds and Schrödinger's first quantum theory. *British Journal for the Philosophy of Science, 62*, 1–27.

Armstrong, D. M. (1961). *Perception and the physical world*. London, UK: Routledge and Kegan Paul.

Barnes, E., & Williams, J. R. G. (2011). A theory of metaphysical indeterminacy. *Oxford Studies in Metaphysics, 6*, 103–148.

Barrett, J. A. (1999). *The quantum mechanics of minds and worlds*. Oxford, UK: Oxford University Press.

Barrett, J. A. (2000). The persistence of memory: Surreal trajectories in Bohm's theory. *Philosophy of Science, 67*, 680–703.

Barrett, J. A. (2003). Are our best physical theories (probably and/or approximately) true? *Philosophy of Science, 70*, 1206–1218.

Barrett, J. A. (2008). Approximate truth and descriptive nesting. *Erkenntnis, 68*, 213–224.

Barrett, J. A. (2011). Everett's pure wave mechanics and the notion of worlds. *European Journal for Philosophy of Science, 1*, 277–302.

Barrett, J. A. (2015). Pure wave mechanics and the very idea of empirical adequacy. *Synthese, 192*, 3071–3104.

Bassi, A., & Ghirardi, G. C. (1999). More about dynamical reduction and the enumeration principle. *British Journal for the Philosophy of Science, 50*, 719–734.

Baym, G. (1969). *Lectures on quantum mechanics*. Redwood City, CA: Addison-Wesley.

Bedard, K. (1999). Material objects in Bohm's interpretation. *Philosophy of Science, 66*, 221–242.

Bell, J. S. (1964). On the Einstein-Podolsky-Rosen paradox. *Physics, 1*, 195–200. Reprinted in Bell (2004), 14–21.

Bell, J. S. (1966). On the problem of hidden variables in quantum mechanics. *Reviews of Modern Physics, 38*, 447–452. Reprinted in Bell (2004), 1–13.

Bell, J. S. (1976a). How to teach special relativity. *Progress in Scientific Culture, 1*, 135–148. Reprinted in Bell (2004), 67–80.

Bell, J. S. (1976b). The measurement theory of Everett and Broglie's pilot wave. In Z. Maric, A. Milojevic, D. Sternheimer, J. P. Vigier, & M. Flato (Eds.), *Quantum mechanics, determinism, causality, and particles* (pp. 11–17). Dordrecht, The Netherlands: Reidel. Reprinted in Bell (2004), 93–99.

Bell, J. S. (1981). Quantum mechanics for cosmologists. In C. Isham, R. Penrose, & D. Sciama (Eds.), *Quantum gravity 2* (pp. 611–637). Oxford, UK: Clarendon Press. Reprinted in Bell (2004), 117–138.

Bell, J. S. (1982). On the impossible pilot wave. *Foundations of Physics, 12*, 989–999. Reprinted in Bell (2004), 159–168.

Bell, J. S. (1987a). Beables for quantum field theory. In B. J. Hiley & F. D. Peat (Eds.), *Quantum implications: Essays in honour of David Bohm* (pp. 227–234). Oxford, UK: Routledge. Reprinted in Bell (2004), 173–180.

Bell, J. S. (1987b). Are there quantum jumps? In C. W. Kilmister (Ed.), *Schrödinger: Centenary celebration of a polymath* (pp. 41–52). Cambridge, UK: Cambridge University Press. Reprinted in Bell (2004), 201–212.

Bell, J. S. (1990). Against "measurement." In A. I. Miller (Ed.), *Sixty-two years of uncertainty: Historical, philosophical, and physical inquiries into the foundations of quantum mechanics* (pp. 17–31). New York, NY: Plenum. Reprinted in Bell (2004), 213–231.

Bell, J. S. (2004). *Speakable and unspeakable in quantum mechanics* (2nd ed.). Cambridge, UK: Cambridge University Press.

Berkovitz, J. (2008). On predictions in retro-causal interpretations of quantum mechanics. *Studies in History and Philosophy of Modern Physics, 39*, 709–735.

Bohm, D. (1951). *Quantum theory*. New York, NY: Prentice-Hall.

Bohm, D. (1952). A suggested interpretation of the quantum theory in terms of "hidden variables" parts I and II. *Physical Review, 85*, 166–179, 180–193. Reprinted in Wheeler and Zurek (1983), 369–396.

Bohm, D. (1957). *Causality and chance in modern physics*. London, UK: Routledge.

Bohn, E. D. (2012). Monism, emergence, and plural logic. *Erkenntnis, 76*, 211–223.

Bohr, N. (1935). Can quantum-mechanical description of physical reality be considered complete? *Physical Review, 48*, 696–702. Reprinted in Wheeler and Zurek (1983), 145–151.

Born, M. (1926). Zur Quantenmechanik der Stoßvorgänge. *Zeitschrift für Physik A, 37*, 863–867. Reprinted and translated as "On the Quantum Mechanics of Collisions" in Wheeler and Zurek (1983), 52–55.

Born, M. (1971). *The Born-Einstein letters*. London, UK: Macmillan.

Born, M., & P. Jordan (1925). Zur Quantenmechanik. *Zeitschrift für Physik A, 34*, 858–888.

Bourget, D., & Chalmers, D. J. (2014). What do philosophers believe? *Philosophical Studies, 170*, 465–500.

Boyd, R. N. (1973). Realism, underdetermination, and a causal theory of evidence. *Noûs, 7*, 1–12.

Bradley, D. J. (2011). Confirmation in a branching world: The Everett interpretation and Sleeping Beauty. *British Journal for the Philosophy of Science, 62*, 323–342.

Brown, H. R., Dewdney, C., & Horton, G. (1995). Bohm particles and their detection in the light of neutron interferometry. *Foundations of Physics, 25*, 329–347.

Callender, C. (2010). The redundancy argument against Bohm's theory. Retrieved November 2015, from http://philosophyfaculty.ucsd.edu/faculty/ccallender/The%20Redundancy%20Argument%20Against%20Bohmian%20Mechanics.doc

Callender, C. (2015). One world, one beable. *Synthese, 192*, 3153–3177.

Calosi, C. (2014). Quantum mechanics and priority monism. *Synthese, 191*, 915–928.

Chalmers, D. J. (1996). *The conscious mind*. Oxford, UK: Oxford University Press.

Clifton, R., & Monton, B. (1999). Losing your marbles in wavefunction collapse theories. *British Journal for the Philosophy of Science, 50*, 697–717.

Cordero, A. (1999). Are GRW tails as bad as they say? *Philosophy of Science, 66*, S59–S71.

Cramer, J. G. (1986). The transactional interpretation of quantum mechanics. *Reviews of Modern Physics, 58*, 647–687.

Davidson, D. (1973). Radical interpretation. *Dialectica, 27*, 314–328.

Davies, P. C. W., & Brown, J. R. (1986). *The ghost in the atom*. Cambridge, UK: Cambridge University Press.

de Broglie, L. (1924). A tentative theory of light quanta. *Philosophical Magazine, 47*, 446–458.

Deutsch, D. (1996). Comment on Lockwood. *British Journal for the Philosophy of Science, 47*, 222–228.

Deutsch, D. (1999). Quantum theory of probability and decisions. *Proceedings of the Royal Society of London A, 455*, 3129–3137.

DeWitt, B. S. (1970). Quantum mechanics and reality. *Physics Today, 23*(9), 30–35. Reprinted in DeWitt and Graham (1973), 155–165.

DeWitt, B. S., & Graham, N. (Eds.). (1973). *The many-worlds interpretation of quantum mechanics.* Princeton, NJ: Princeton University Press.

Dickson, M. (1995). An empirical reply to empiricism: Protective measurement opens the door for quantum realism. *Philosophy of Science, 62,* 122–140.

Dürr, D., Goldstein, S., & Zanghì, N. (1997). Bohmian mechanics and the meaning of the wave function. In R. S. Cohen, M. Horne, & J. Stachel (Eds.), *Experimental metaphysics: Quantum mechanical studies for Abner Shimony, Volume One.* (Boston Studies in the Philosophy of Science, Vol. 193, pp. 25–38). Dordrecht, The Netherlands: Kluwer.

Einstein, A., Podolsky, B., & Rosen, N. (1935). Can quantum-mechanical description of physical reality be considered complete? *Physical Review, 47,* 777–780. Reprinted in Wheeler and Zurek (1983), 138–141.

Englert, B.-G., Scully, M. O., Süssmann, G., & Walther, H. (1992). Surrealistic Bohm trajectories. *Zeitschrift für Naturforschung, 47a,* 1175–1186.

Everett, H., III. (1957). "Relative state" formulation of quantum mechanics. *Reviews of Modern Physics, 29,* 454–462. Reprinted in DeWitt and Graham (1973), 141–149, and in Wheeler and Zurek (1983), 315–323.

Everett, H., III., Barrett, J. A., & Byrne, P. (2012). *The Everett interpretation of quantum mechanics: Collected works 1955–1980 with commentary.* Princeton, NJ: Princeton University Press.

Faye, J. (2010). Niels Bohr and the Vienna Circle. In J. Manninen & F. Stadler (Eds.), *The Vienna Circle in the Nordic countries* (The Vienna Circle Institute Yearbook 14, pp. 33–45). Dordrecht, The Netherlands: Springer.

French, S. (1989). Identity and individuality in classical and quantum physics. *Australasian Journal of Philosophy, 67,* 432–446.

French, S., & Krause, D. (2006). *Identity in physics: A historical, philosophical, and formal analysis.* Oxford, UK: Clarendon Press.

French, S., and Ladyman, J. (2003). Remodelling structural realism: Quantum physics and the metaphysics of structure. *Synthese, 136,* 31–56.

Frigg, R. (2003). On the property structure of realist collapse interpretations of quantum mechanics and the so-called "counting anomaly." *International Studies in the Philosophy of Science, 17,* 43–57.

Frigg, R., & Votsis, I. (2011). Everything you always wanted to know about structural realism but were afraid to ask. *European Journal for Philosophy of Science, 1,* 227–276.

Gamow, G. (1966). *Thirty years that shook physics: The story of quantum theory.* New York, NY: Doubleday.

Ghirardi, G. C., Grassi, R., & Benatti, F. (1995). Describing the macroscopic world: Closing the circle within the dynamical reduction program. *Foundations of Physics, 25,* 5–38.

Ghirardi, G. C., Pearle, P., and Rimini, A. (1990). Markov processes in Hilbert space and continuous spontaneous localization of systems of identical particles. *Physical Review A, 42,* 78–89.

Ghirardi, G. C., Rimini, A., & Weber, T. (1986). Unified dynamics for microscopic and macroscopic systems. *Physical Review D, 34,* 470–491.

Goldstein, S., & Zanghì, N. (2013). Reality and the role of the wave function in quantum theory. In A. Ney & D. Z. Albert (Eds.), *The wave function* (pp. 91–109). Oxford, UK: Oxford University Press.

Greaves, H. (2004). Understanding Deutsch's probability in a deterministic multiverse. *Studies in History and Philosophy of Modern Physics, 35*, 423–456.

Grice, P. (1957). Meaning. *Philosophical Review, 66*, 377–388.

Groisman, B., Hallakoun, N., & Vaidman, L. (2013). The measure of existence of a quantum world and the Sleeping Beauty problem. *Analysis, 73*, 695–706.

Halvorson, H., and Clifton, R. (2002). Reconsidering Bohr's reply to EPR. In T. Placek & J. Butterfield (Eds.), *Non-locality and modality* (pp. 3–18). Dordrecht, The Netherlands: Kluwer.

Hawthorne, J., & Silberstein, M. (1995). For whom the Bell arguments toll. *Synthese, 102*, 99–138.

Heisenberg, W. (1925). Über quantentheoretische Umdeutung kinematischer und mechanischer Beziehungen. *Zeitschrift für Physik A, 33*, 879–893.

Howard, D. (1994). What makes a classical concept classical? In J. Faye & H. Folse (Eds.), *Niels Bohr and contemporary philosophy* (pp. 201–229). Dordrecht, The Netherlands: Springer.

Howard, D. (2004). Who invented the "Copenhagen interpretation"? A study in mythology. *Philosophy of Science, 71*, 669–682.

Huggett, N. (1999). *Space from Zeno to Einstein*. Cambridge, MA: MIT Press.

Hyde, D. (1994). Why higher-order vagueness is a pseudo-problem. *Mind, 103*, 35–41.

Ismael, J. (2003). How to combine chance and determinism: Thinking about the future in an Everett universe. *Philosophy of Science, 70*, 776–790.

Johnson, W. E. (1921). *Logic, part I*. Cambridge, UK: Cambridge University Press.

Kane, R. (1996). *The significance of free will*. New York, NY: Oxford University Press.

Kastner, R. E. (2013). *The transactional interpretation of quantum mechanics: The reality of possibility*. Cambridge, UK: Cambridge University Press.

Kent, A. (2010). One world versus many: The inadequacy of Everettian accounts of evolution, probability, and scientific confirmation. In S. Saunders, J. Barrett, A. Kent, & D. Wallace (Eds.), *Many worlds: Everett, quantum theory, and reality* (pp. 307–354). Oxford, UK: Oxford University Press.

Kochen, S., & Specker, E. (1967). The problem of hidden variables in quantum mechanics. *Journal of Mathematics and Mechanics, 17*, 59–87.

Kripke, S. A. (1980). *Naming and necessity*. Cambridge, MA: Harvard University Press.

Ladyman, J. (1998). What is structural realism? *Studies in History and Philosophy of Science, 29*, 409–424.

Ladyman, J., & Ross, D., (2007). *Everything must go: Metaphysics naturalized*. Oxford, UK: Oxford University Press.

Laudan, L. (1981). A confutation of convergent realism. *Philosophy of Science, 48*, 19–49.

Lewis, D. (1976). Survival and identity. In A. K. Rorty (Ed.), *The identities of persons* (pp. 17–40). Berkeley: University of California Press.

Lewis, D. (1979). Attitudes de dicto and de se. *Philosophical Review, 88*, 513–543.

Lewis, D. (1986a). *On the plurality of worlds*. Oxford, UK: Blackwell.
Lewis, D. (1986b). *Philosophical papers* (Vol. 2). Oxford, UK: Oxford University Press.
Lewis, D. (1993). Many, but almost one. In K. Campbell, J. Bacon, & L. Reinhardt (Eds.), *Ontology, causality, and mind: Essays on the philosophy of D. M. Armstrong* (pp. 23–38). Cambridge, UK: Cambridge University Press.
Lewis, D. (2004). How many lives has Schrödinger's cat? *Australasian Journal of Philosophy, 82*, 3–22.
Lewis, P. J. (1997). Quantum mechanics, orthogonality, and counting. *British Journal for the Philosophy of Science, 48*, 313–328.
Lewis, P. J. (2000). What is it like to be Schrödinger's cat? *Analysis, 60*, 22–29.
Lewis, P. J. (2003). Four strategies for dealing with the counting anomaly in spontaneous collapse theories of quantum mechanics. *International Studies in Philosophy of Science, 17*, 137–142.
Lewis, P. J. (2004). Life in configuration space. *British Journal for the Philosophy of Science, 55*, 713–729.
Lewis, P. J. (2006a). Conspiracy theories of quantum mechanics. *British Journal for the Philosophy of Science, 57*, 359–381.
Lewis, P. J. (2006b). GRW: A case study in quantum ontology. *Philosophy Compass, 1*, 224–244.
Lewis, P. J. (2007a). Quantum Sleeping Beauty. *Analysis, 67*, 59–65.
Lewis, P. J. (2007b). Uncertainty and probability for branching selves. *Studies in History and Philosophy of Modern Physics, 38*, 1–14.
Lewis, P. J. (2009). Probability, self-location, and quantum branching. *Philosophy of Science, 76*, 1009–1019.
Lewis, P. J, (2010). Probability in Everettian quantum mechanics. *Manuscrito, 33*, 285–306.
Lewis, P. J. (2013a). Dimension and illusion. In A. Ney & D. Z. Albert (Eds.), *The wave function* (pp. 110–125). Oxford, UK: Oxford University Press.
Lewis, P. J. (2013b). Retrocausal quantum mechanics: Maudlin's challenge revisited. *Studies in History and Philosophy of Modern Physics, 44*, 442–449.
Lockwood, M. (1989). *Mind, brain and the quantum: The compound 'I.'*. Oxford, UK: Blackwell.
Loewer, B. (1996). Humean supervenience. *Philosophical Topics, 24*, 101–127.
Makinson, D. C. (1965). The paradox of the preface. *Analysis, 25*, 205–207.
Maudlin, T. (1994). *Quantum non-locality and relativity*. Oxford, UK: Blackwell.
Maudlin, T. (2013). The nature of the quantum state. In A. Ney & D. Z. Albert (Eds.), *The wave function* (pp. 126–153). Oxford, UK: Oxford University Press.
Mermin, N. D. (1981). Quantum mysteries for anyone. *Journal of Philosophy, 78*, 397–408.
Mermin, N. D. (1993). Hidden variables and the two theorems of John Bell. *Reviews of Modern Physics, 65*, 803–815.
Mermin, N. D. (1998). What is quantum mechanics trying to tell us? *American Journal of Physics, 66*, 753–767.
Merricks, T. (2001). *Objects and persons*. Oxford, UK: Clarendon Press.
Monton, B. (2002). Wavefunction ontology. *Synthese, 130*, 265–277.
Monton, B. (2006). Quantum mechanics and 3N-dimensional space. *Philosophy of Science, 73*, 778–789.

Monton, B. (2013). Against 3N-dimensional space. In A. Ney & D. Z. Albert (Eds.), *The wave function* (pp. 154–167). Oxford, UK: Oxford University Press.

Myrvold, W. C. (2002). On peaceful coexistence: Is the collapse postulate incompatible with relativity? *Studies in History and Philosophy of Modern Physics, 33*, 435–466.

Myrvold, W. C. (2015). What is a wavefunction? *Synthese, 192*, 3247–3274.

Ney, A. (2013). Ontological reduction and the wave function ontology. In A. Ney & D. Z. Albert (Eds.), *The wave function* (pp. 168–183). Oxford, UK: Oxford University Press.

Ney, A., & Albert, D. Z. (Eds.). (2013). *The wave function*. Oxford, UK: Oxford University Press.

North, J. (2013). The structure of a quantum world. In A. Ney & D. Z. Albert (Eds.), *The wave function* (pp. 184–202). Oxford, UK: Oxford University Press.

Okasha, S. (2002). Darwinian metaphysics: Species and the question of essentialism. *Synthese, 131*, 191–213.

Papineau, D. (2004). David Lewis and Schrödinger's cat. *Australasian Journal of Philosophy, 82*, 153–169.

Parfit, D. (1971). Personal identity. *Philosophical Review, 80*, 3–27.

Pearle, P. (1976). Reduction of the state vector by a nonlinear Schrödinger equation. *Physical Review D, 13*, 857–868.

Perry, J. (1979). The problem of the essential indexical. *Noûs, 13*, 3–21.

Philippidis, C., Bohm, D., & Kaye, R. D. (1982). The Aharonov-Bohm effect and the quantum potential. *Il Nuovo Cimento B, 71*, 75–88.

Polkinghorne, J. (2002). *Quantum theory: A very short introduction*. Oxford, UK: Oxford University Press.

Price, H. (1994). A neglected route to realism about quantum mechanics. *Mind, 103*, 303–336.

Price, H. (1996). *Time's arrow and Archimedes' point*. Oxford, UK: Oxford University Press.

Price, H. (2010). Decisions, decisions, decisions: Can Savage salvage Everettian probability? In S. Saunders, J. Barrett, A. Kent, & D. Wallace (Eds.), *Many worlds: Everett, quantum theory, and reality* (pp. 369–390). Oxford, UK: Oxford University Press.

Price, H. (2012). Does time-symmetry imply retrocausality? How the quantum world says "maybe?" *Studies in History and Philosophy of Modern Physics, 43*, 75–83.

Psillos, S. (1999). *Scientific realism: How science tracks truth*. London, UK: Routledge.

Rovelli, C. (1996). Relational quantum mechanics. *International Journal of Theoretical Physics, 35*, 1637–1678.

Sanford, D. H. (2011). Determinates vs. determinables. In E. N. Zalta (Ed.), *The Stanford encyclopedia of philosophy*. Retrieved October 2015, from http://plato.stanford.edu/archives/spr2011/entries/determinate-determinables/

Saunders, S. (1995). Time, quantum mechanics, and decoherence. *Synthese, 102*, 235–266.

Saunders, S., & Wallace, D. (2008). Branching and uncertainty. *British Journal for the Philosophy of Science, 59*, 293–305.

Saunders, S., Barrett, J., Kent, A., & Wallace, D. (Eds.), (2010). *Many worlds: Everett, quantum theory, and reality*. Oxford, UK: Oxford University Press.

Savage, L. J. (1954). *The foundations of statistics*. New York, NY: Wiley.
Schaffer, J. (2010). Monism: The priority of the whole. *Philosophical Review, 119,* 31–76.
Schrödinger, E. (1926a). An undulatory theory of the mechanics of atoms and molecules. *Physical Review, 28,* 1049–1070.
Schrödinger, E. (1926b). Quantisierung als Eigenwertproblem. *Annalen der Physik, 384,* 489–527.
Schrödinger, E. (1935). Die gegenwärtige Situation in der Quantenmechanik. *Naturwissenschaften, 23,* 807–812, 823–828, 844–849. Reprinted and translated as "The present situation in quantum mechanics." In J. A. Wheeler & W. H. Zurek (Eds.). (1983) *Quantum theory and measurement* (pp. 152–167). Princeton, NJ: Princeton University Press.
Schwitzgebel, E. (2008). The unreliability of naive introspection. *Philosophical Review, 117,* 245–273.
Sider, T. (1996). All the world's a stage. *Australasian Journal of Philosophy, 74,* 433–453.
Sider, T. (2014). *Asymmetric personal identity*. Retrieved December 2015, from http://tedsider.org/papers/asymmetric_personal_identity.pdf
Silberstein, M. (2015). *Quantum mechanics*. Retrieved December 2015, from http://philpapers.org/rec/SILQM
Sutherland, R. I. (2008). Causally symmetric Bohm model. *Studies in History and Philosophy of Modern Physics, 39,* 782–805.
Tappenden, P. (2008). Saunders and Wallace on Everett and Lewis. *British Journal for the Philosophy of Science, 59,* 307–314.
Taylor, R. (1962). Fatalism. *Philosophical Review, 71,* 56–66.
Tegmark, M. (1998). The interpretation of quantum mechanics: Many worlds or many words? *Fortschritte der Physik, 46,* 855–862.
Teller, P. (1986). Relational holism and quantum mechanics. *British Journal for the Philosophy of Science, 37,* 71–81.
Tumulka, R. (2006a). A relativistic version of the Ghirardi-Rimini-Weber model. *Journal of Statistical Physics, 125,* 821–840.
Tumulka, R. (2006b). Collapse and relativity. In A. Bassi, D. Dürr, T. Weber, & N. Zanghì (Eds.), Quantum mechanics: Are there quantum jumps? and On the Present Status of Quantum Mechanics. *AIP Conference Proceedings, 844,* 340–351.
Tye, M. (2000). Vagueness and reality. *Philosophical Topics, 28,* 195–209.
Vaidman, L. (1998). On schizophrenic experiences of the neutron or why we should believe in the many-worlds interpretation of quantum theory. *International Studies in the Philosophy of Science, 12,* 245–261.
Vaidman, L. (2010). Time symmetry and the many-worlds interpretation. In S. Saunders, J. Barrett, A. Kent, & D. Wallace (Eds.), *Many worlds: Everett, quantum theory, and reality* (pp. 582–596). Oxford, UK: Oxford University Press.
van Fraassen, B. C.(1979). Hidden variables and the modal interpretation of quantum theory. *Synthese, 42,* 155–165.
van Fraassen, B. C. (1980). *The scientific image*. Oxford, UK: Oxford University Press.
van Fraassen, B. C. (1991). *Quantum mechanics: An empiricist view*. Oxford, UK: Clarendon Press.
Vink, J. C. (1993). Quantum mechanics in terms of discrete beables. *Physical Review A, 48,* 1808–1818.

von Neumann, J. (1932). *Mathematische Grundlagen der Quantenmechanik*. Berlin, Germany: Springer. Translated and reprinted (1955) as *Mathematical Foundations of Quantum Mechanics*. Princeton, NJ: Princeton University Press.

Wallace, D. (2003a). Everett and structure. *Studies in History and Philosophy of Modern Physics, 34*, 87–105.

Wallace, D. (2003b). Everettian rationality: Defending Deutsch's approach to probability in the Everett interpretation. *Studies in History and Philosophy of Modern Physics, 34*, 415–439.

Wallace, D. (2007). Quantum probability from subjective likelihood: Improving on Deutsch's proof of the probability rule. *Studies in History and Philosophy of Modern Physics, 38*, 311–332.

Wallace, D. (2010). Decoherence and ontology. In S. Saunders, J. Barrett, A. Kent, & D. Wallace (Eds.), *Many worlds: Everett, quantum theory, and reality* (pp. 53–72). Oxford, UK: Oxford University Press.

Wallace, D. (2012). *The emergent multiverse*. Oxford, UK: Oxford University Press.

Wallace, D. (2014). *Life and death in the tails of the GRW wave function*. Retrieved October 2015, from http://arxiv.org/abs/1407.4746

Wallace, D., & Timpson, C. G. (2010). Quantum mechanics on spacetime I: Spacetime state realism. *British Journal for the Philosophy of Science, 61*, 697–727.

Wharton, K. (2010). Time-symmetric boundary conditions and quantum foundations. *Symmetry, 2*, 272–283.

Wheeler, J. A., & Zurek, W. H. (Eds.). (1983) *Quantum theory and measurement*. Princeton, NJ: Princeton University Press.

Williamson, T. (1994). *Vagueness*. London, UK: Routledge.

Zeh, H. D. (1999). Why Bohm's quantum theory? *Foundations of Physics Letters, 12*, 197–200.

Zurek, W. H. (1991). Decoherence and the transition from quantum to classical. *Physics Today, 44*(10), 36–44.

INDEX

Aharonov, Y., 118
Albert, D. Z., 54, 80–84, 91, 131–132, 154–163, 189, 191
Allori, V., 158–163, 192
appearance. *See* experience
Armstrong, D. M., 73

bare theory, 81–87
Barnes, E., 77
Barrett, J. A., 80–84, 183, 187, 188, 189, 190
Bassi, A., 94–95
Baym, G., 76, 78
Bedard, K., 193
Bell, J. S., 50, 59, 67, 88, 114, 125, 154, 179, 185. *See also* Bell's theorem
Bell's theorem, 9, 59
    and causal nonlocality, 107, 111, 115–116, 123–124
    and free will, 147–149
    and holism, 167–172
    independence assumption of, 36–40, 59, 115, 123, 147, 167–168, 172
    locality assumption of, 36, 42, 59, 111, 167–170
    proof of, 34–37, 185
    and realism, 38, 41–43
Berkovitz, J., 117–118
block universe, 124

Bohm, D., 27, 154. *See also* Bohm's theory
Bohm's theory, 41–42, 55–59, 180, 186
    and causal nonlocality, 111–115, 119–120, 127
    and determinism, 128–130, 145, 146, 149
    and dimensionality, 151, 155, 157, 159–162
    and holism, 169–170
    and indeterminacy, 95, 101–105
    and individuals, 177
    and underdetermination, 64–71
Bohr, N., 30–34, 39, 107, 126, 179, 184, 185
Born, M., 9. *See also* Born rule.
Born rule, 23, 39, 74, 118, 145, 154
    in Bohm's theory, 56–57, 114
    in GRW theory, 51–53
    in many-worlds theory, 63, 69, 129, 135–142, 149
    in matrix mechanics, 13–17
    in wave mechanics, 19–21
brain in a vat, 83–84
brain states, 91–92, 100–101, 123, 131–132, 146, 181
branching indifference, 137–142, 144–145
Brown, H. R., 193

Callender, C., 69, 192
caring measure, 138
causal determinism. *See* determinism
causal explanation, 107–127, 181
causal loop, 117
causation
    backwards. *See* hidden variable theories, retrocausal
    nonlocal. *See* locality, causal
Chalmers, D. J., 193
classical mechanics, xiv–xv, 5, 7, 46, 107–110, 114, 145, 155
Clifton, R., 92, 184
collapse
    effective, 65, 99
    GRW, 51–54, 66, 86–97, 122–125, 146, 156, 181
    postulate, 14, 29, 49, 128
common cause, 40, 116, 148, 168
complementarity, 33, 179
configuration space, 53, 151–164, 171, 177
conspiracy theory, 116–117, 148
contextuality, 32–33, 39
Cordero, A., 66
counting anomaly, 92–96, 101, 103
Cramer, J. G., 123–125

Davidson, D., 85
death, 142–145
de Broglie, L., 5
decision theory, 136–142, 145
decoherence, 97–101
Dennett's criterion, 66–69
determinable property, 72–73, 104–106
determinacy. *See* indeterminacy
determinism, 41, 51, 57, 112, 128–129, 180–182
    and free will, 145–149
Deutsch, D., 66, 136–142
Dewdney, C., 193
DeWitt, B. S., 62, 65, 136, 187
diachronic consistency, 136–137, 140–141
Dickson, M., 185
dimensions. *See* space
Dirac equation, 120
Dirac notation, 10

dualism, 131–132, 179
Dürr, D., 161
dynamical law, 41, 69
    Bohmian, 56–58, 112–114, 128, 157, 180
    GRW. *See* collapse, GRW
    linear, 15, 47
    Schrödinger. *See* Schrödinger, equation

eigenstate, 10–14, 17–18, 28, 54, 76
eigenstate-eigenvalue link. *See* strict link
eigenvalue, 10–11
eigenvector. *See* eigenstate
Einstein, A., 23, 29, 38, 39, 110, 114. *See also* EPR argument
emergence, 63, 97, 166–173
empirical adequacy, 26, 43, 61, 69–71, 81–82, 135, 142
empirical incoherence, 83
entanglement, 7–9, 15–17, 22, 27–36, 149
    and holism, 166–173
    and individuals, 173–174, 177
    and locality, 111–113, 119, 123–124
EPR argument, 27–34, 38–39
Everett, H., 59–65, 187, 188, 191. *See also* many-worlds theory
expectation value, 142–143, 159
expected utility, 137–139
experience, xiv–xv, 61, 80–85, 91, 99–100

free will, 115, 116, 145–149
French, S., 175–176, 193
Frigg, R., 93, 176, 189
functionalism
    about macro-objects. *See* Dennett's criterion
    about mind, 131–132
    about probability, 138
fuzzy link, 87–103

Ghirardi, G. C., 94–95, 192. *See also* GRW theory

Goldstein, S., 161
gravity, 108
Greaves, H., 138
Grice, P., 85
GRW theory, 51–55, 180–181
   and causal nonlocality, 122–123
   and determinism, 128, 145–146
   and dimensionality, 151–152, 156–157, 159–160, 163
   flashy, 55, 125–126, 159–160, 163, 178, 181
   and holism, 169–170
   and indeterminacy, 86–97
   and individuals, 177–178
   massy, 55, 94–95, 159–160, 163, 181
   and underdetermination, 64–71

Halvorson, H., 184
Hawthorne, J., 167–168
Heisenberg, W., 9
hidden variables, 38, 56, 58–59
hidden variable theories, 59, 180
   Bohmian. See Bohm's theory
   local, 115–118, 147–149
   modal, 59
   retrocausal, 59, 104, 116–118, 148–149, 162, 172–173, 177, 180
holism, 165–173
   relational, 166–167
Horton, G., 193
Howard, D., 184

immortality. See death
indeterminacy, 72–106
   brute, 78
   compositional, 73–74
   of experience, 80–85, 91
   quantum, 77–79, 88, 96–97, 101, 103, 106
indexical reference, 133–134
individuals, 173–178
individuation
   of branches, 139
   of minds, 132–133
   of particles. See individuals
   of people, 132–133, 147

instrumentalism, 32
interference, 2–6, 21–22, 52, 55, 57–58, 62, 97–98, 121–122
interpretation, 22–24
Ismael, J., 134

Jordan, P., 9

Kane, R., 145–146
Kastner, R., 123–125
Klein-Gordon equation, 120
Kochen-Specker theorem, 37–42, 103–104
Krause, D., 193
Kripke, S. A., 85

Ladyman, J., 174–176
Leibniz, G., 109–110, 114, 174
Lewis, D., 68, 73–74, 133, 142–144, 165. See also Lewisian people
Lewisian people, 133–136, 147
locality
   assumption. See Bell's theorem, locality assumption of
   causal, 29, 108–127, 169, 180–181
Loewer, B., 131–132, 171

many-minds theory, 132–133
many-worlds theory, 41–42, 59–64, 181–182
   and causal locality, 119–120
   and determinism, 128–135, 146–147
   and dimensionality, 151, 154, 156, 163
   and holism, 168, 170
   and immortality, 142–145
   and indeterminacy, 97–101
   and individuals, 177–178
   massy, 159
   and probability, 135–142
   and uncertainty, 129–135
   and underdetermination, 64–71
mass density distribution, 94, 159, 181
matrix mechanics, 10–17
Maudlin, T., 114, 125, 158–163, 190, 192
meaning, 23, 85–86

measurement, 13–14, 19–20, 29–42, 78, 91
  postulate. *See* collapse, postulate
  problem, 45–50, 53, 57, 59–63
Mermin, N. D., 176, 185
Merricks, T., xv, 173
modal theories, 59
Monton, B., 92–93, 157–158, 162–163
Myrvold, W. C., 125, 171

Newton, I., 23, 109, 114, 115
Newtonian mechanics. *See* classical mechanics
Ney, A., 187
nonlocality. *See* locality
North, J., 160, 192

ontic structural realism, 175–176
operator, 10–13
  Hamiltonian, 15, 21, 155–157

Parfit, D., 63, 133
Pearle, P., 51
Perry, J., 190
personal identity, 63, 133, 144
phenomena, xiv–xviii, 1–9, 22–24
physical state (*vs.* quantum state), 9–10
Podolsky, B. *See* EPR argument
preface paradox, 93–94
Price, H., 40, 104, 116–118, 191
primitive ontology, 159–161, 181
principle of the identity of indiscernibles, 174–175
priority monism, 173–174
probability, 135–142
projection, 13–14
  postulate. *See* collapse, postulate

quantization, 7
quantum Russian roulette, 142
quantum state (*vs.* physical state), 9–10

rationality, 136–142
realism, 25–26, 38, 40–43, 44–45
  structural. *See* ontic structural realism

reductionism, part-whole. *See* supervenience, Humean
relational quantum mechanics, 175–176
relations, 166–170
  as fundamental, 175–176
relative state, 60–61
representation theorem, 137–138
retrocausal theories
  hidden variable. *See* hidden variable theories, retrocausal
  spontaneous collapse. *See* transactional interpretation
Rimini, A. *See* GRW theory
Rosen, N. *See* EPR argument
Ross, D., 174–176
Rovelli, C., 175

Sanford, D. H., 74
Saunders, S., 63, 133, 187
Schaffer, J., 173–174, 193
Schrödinger, 9, 23, 154
  cat, 49–50, 65–69, 87
  equation, 15, 20–21, 49, 119–120, 155
Schwitzgebel, E., 188
self-location, 121, 190–191
semantic incoherence, 85–86
Sider, T., 133–134
Silberstein, M., 167–168, 192
simultaneity, 110–111, 114, 123, 126
single-mind theory, 131–132
skepticism, 83–84
solipsism, 100
space
  three-dimensional, 152–163
  $3N$-dimensional. *See* configuration space
special relativity, 109–115, 120, 125–126
spin, 7–8
spontaneous collapse theories, 41, 50–51, 55, 123–126, 180–181. *See also* GRW theory
Star Trek transporter, 143–144
strict link, 76–77, 79, 87, 94, 99
string theory, 157
superposition, 12

supervenience,
   Humean, 165–173
   of the mental on the physical, 129–132
Sutherland, R. I., 118

Teller, P., 166–167
theoretical virtues, 44–45
Timpson, C. G., 162, 190, 193
transactional interpretation, 123–125
Tumulka, R., 125–126, 159
Tye, M., 73–74, 77

uncertainty, 129–134, 138, 144
underdetermination, 44–45, 64–71, 175–176

vague link, 89–90, 95–97, 100
vagueness, 73, 77, 79, 86–90, 95–97, 106
Vaidman, L., 91, 118, 131
van Fraassen, B., 25–26, 44, 59, 186

vector, 10
vector space, 11–12
Vink, J. C., 59
von Neumann, J., 64, 185

Wallace, D., 63–64, 66, 120–121, 133, 136–142, 144, 162–163
wave function, 18–23, 55, 58–59, 63, 154–164
   as epistemic, 161–162
   as a law, 161
wave mechanics, 17–22
wave-particle duality, 6, 55, 63
Weber, T. *See* GRW theory
Wharton, K., 118
Wigner, E., 179
Williams, J. R. G., 77
Williamson, T., 188

Zanghì, N., 161
Zeh, H. D., 65–66
Zurek, W. H., 189